FOOD FROM DRY LANDS

Systems Approaches for Sustainable Agricultural Development

Volume 1

The titles published in this series are listed at the end of this volume.

Food from dry lands

An integrated approach to planning of
agricultural development

Edited by

Th. ALBERDA

Formerly DLO – Centre for
Agrobiological Research (CABO-DLO),
Wageningen, The Netherlands

H. van KEULEN

DLO – Centre for Agrobiological
Research (CABO-DLO), Wageningen,
The Netherlands

N.G. SELIGMAN

Agricultural Research Organization
(ARO), Bet Dagan, Israel

C.T. de WIT

Emeritus Professor Wageningen
Agricultural University (WAU),
Wageningen, The Netherlands

Springer-Science+Business Media, B.V.

Library of Congress Cataloging-in-Publication Data

Food from dry lands : an integrated approach to planning of
   agricultural development / edited by Th. Alberda ... [et al.].
       p.   cm. -- (System approaches for sustainable agricultural
   development ; v. 1)
   Includes bibliographical references.
   ISBN 978-0-7923-1877-4   ISBN 978-94-011-2830-8 (eBook)
   DOI 10.1007/978-94-011-2830-8
   1. Arid regions agriculture. 2. Agropastoral systems.
   3. Pastures. 4. Sheep.   I. Alberda, Th. II. Series.
   S613.F57   1992
   338.1'763'009154--dc20   92-21701

ISBN 978-0-7923-1877-4

printed on acid-free paper

Contents

Contributors

Alberda, Th.	formerly DLO – Centre for Agrobiological Research (CABO-DLO), Wageningen, The Netherlands.
Benjamin, R.W.	Agricultural Research Organisation, Bet Dagan, Israel.
Keulen, H. van	DLO – Centre for Agrobiological Research (CABO-DLO), Wageningen, The Netherlands.
Noy-Meir, I.	Department of Evolution, Systematics and Ecology, Hebrew University of Jerusalem, Jerusalem, Israel.
Seligman, N.G.	Agricultural Research Organization (ARO), Bet Dagan, Israel.
Spharim, I.	Agricultural Research Organization, Bet Dagan, Israel.
Spharim, R.	Bahat – Examination and Evaluation of Technologies Ltd., Jerusalem, Israel.
Ungar, E.D.	Agricultural Research Organization, Bet Dagan, Israel.
Wit, C.T. de	Emeritus Professor Wageningen Agricultural University (WAU), Wageningen, The Netherlands.

Foreword

In the early seventies, scientists in Israel and The Netherlands started a cooperative project on actual and potential production under semi-arid conditions. In Israel research concentrated on primary production of natural pastures and small grain crops, and on the associated secondary production of small ruminants. Most of the experimental work was carried out at the Migda Experimental Farm in the semi-desert of the northern Negev where the long-term average annual rainfall is 250 mm. In The Netherlands existing facilities in Wageningen were used for measuring growth, photosynthesis and transpiration of Negev pasture plants and small grains under controlled conditions, both as individual plants and as simulated swards and crops.

The joint research program was initiated by the late N.H. Tadmor and A. Dovrat from Israel and by C.T. de Wit and Th. Alberda from The Netherlands and conducted by various scientists from both countries, some of whom are among the authors of this book.

The experimental results first served as a basis for the development, calibration and validation of simulation models of the growth and water use of pasture and crops. Subsequently, additional models were developed, allowing incorporation of socio-economic considerations, both at the farm and regional level, so harnessing the research results for analysis of regional development possibilities.

The methodologies developed within the present project served as a basis for the initiation of projects in Mali, Egypt and Peru. Although none of these projects were as all-encompassing as that in Israel, they have demonstrated the possibilities of applying the tools developed in one region for extrapolation and prediction in others, provided that the relevant input parameters for the local conditions are available.

This book presents the results of the Netherlands/Israeli project, with emphasis on the methodology for exploring different regional development options. It starts with a summary of the experimental results on primary and secondary production and illustrates how these results can be integrated in simulation programs and optimization techniques as a basis for development planning of a farm or of a region. To improve readability, the models have not been included in the text, but are given in appendices or in relevant references. The book is intended for biologists, agricultural scientists and planners involved in the development planning, particularly of semi-arid regions. It describes a method for estimating the socio-economic effectiveness of technological innovation in a realistic multiple-goal context. This method can provide 'early warning' of pitfalls in development schemes and so improve development planning in a world where this is becoming more complicated by the day.

Acknowledgements

The editors wish to thank:
- the authors, not only for writing their chapters, but also for their continuous efforts in realizing the book in the form it is presented;
- the directors of the participating research institutes: the Agricultural Research Organisation (ARO), Bet Dagan, Israel, and the Centre for Agrobiological Research (CABO-DLO), Wageningen, The Netherlands;
- the Vice-Chancellor of the Hebrew University of Jerusalem, Jerusalem, Israel and the Rector of the Agricultural University Wageningen (WAU), Wageningen, The Netherlands;
- the Directorate General of International Cooperation (DGIS) of the Dutch Ministry of Foreign Affairs, The Hague, The Netherlands, for financial support to conduct both the research and the preparation of this book.

Thanks are also due to the following persons, who participated in the research, but, for various reasons, did not contribute to this book: E. Dayan, A. Dovrat, Y. Harpaz, R. Jonathan, H. Lof, G. Orshan and the late N.H. Tadmor.

1. Introduction

C.T. DE WIT and N.G. SELIGMAN

1.1. In the current of change

The ongoing industrial revolution and technological achievements of agriculture in the western world during the last half century have upset basic concepts of man's priorities. From a major element in the economy of nations, agriculture declined to an adjunct of an ever-expanding industrial and service economy, while the distance between the primary producer and the consumer has widened to the extent that agriculture has almost faded from the consciousness of the majority of the population.

This development has also affected the semi-arid regions in West Asia and North Africa, where large areas are close to urban markets and to sources of industrial inputs. This logistic proximity creates favourable conditions for transport of inputs and outputs across the borders of the agricultural sector. For traditional agriculture this advantage is reduced by the fact that production and prices in the region are exposed to competition from other parts of the world, more lavishly endowed with natural agricultural resources. In addition, input costs, particularly labour costs, increase in the wake of industrialization to a level that makes many traditional agricultural technologies obsolete. Research under such conditions cannot restrict itself to understanding and supporting incremental changes in the agricultural system, but has to consider the whole spectrum of technological options for the region to identify those that meet as far as possible the aspirations of the main stakeholders. This requires innovative dovetailing of agro-ecological and socio-economic research so as to widen the scope for managerial and development manipulation.

The present volume aims to contribute to the development of such an approach based on the semi-arid region of Israel, an area that has been intensively exposed to technological, socio-economic and political change over the past half century. The method employed, however, is not site specific and can be applied more widely to understand and guide development in other parts of the mediterranean region that are also being swept up in the currents of change.

Th. Alberda et al. (eds.), Food from Dry Lands, 1–5.

1.2. The natural environment

The semi-arid zones with a mediterranean type of climate are characterized by a hot dry summer and a mild to cool rainy winter. They are situated around the eastern part of the Mediterranean, especially in the Middle East, in the south of Greece, Italy and Spain and also in North Africa and Iran. Smaller areas occur in California, central Chile, the south-western corner of South Africa and in the south and southwest of Australia.

The natural vegetation that dominates these semi-arid regions in the Mediterranean Basin is a dwarf shrub steppe interspersed with herbaceous species, mainly annuals. In much of the area, trees, taller shrubs and even dwarf shrubs have been cut for fuel and building material, leaving annual and perennial herbs as the dominating plant forms. High seed yields and seed longevity allow the annuals to persist, even under heavy grazing pressure, but the period that they remain green is relatively short. When undergrazed, much of the production of shrubs goes to inedible woody tissues. Under heavy grazing most newly developing and protuberant twigs are eaten. Leaf development and water use in the rainy season are then limited, so that the water availability may extend well into the dry season. The dense and woody structure of these dwarf shrubs provides protection against destruction by occasional overgrazing.

Small grains may be cultivated on the better soils, but pastoralism is the most common form of exploitation. The most common domestic animals in the mediterranean semi-arid regions are sheep and goats. Because their short reproductive cycle coincides more or less with the green season, they are less affected by seasonal food shortages than cattle and, therefore, are suitable not only for transhumance but are also well adapted to sedentary farming systems.

Much of the field research discussed in this book was carried out at the Migda Experimental Farm of the Agricultural Research Organization in the semi-arid northern Negev of Israel where average annual rainfall is 250 mm and the vegetation is dominated by native winter annuals. After the first effective rains between October and December, germination takes one to two weeks, depending on temperature. Lambing occurs at the beginning of the grand period of vegetative growth in January and February. Standing biomass reaches a peak at the beginning of April when the seeds ripen. Then the vegetation dries up and even without grazing gradually diminishes through weathering by wind and decomposition. The variability in annual rainfall causes substantial variability in quality and quantity of the available biomass. In good rainfall years the quantity of biomass may be larger, but the quality may be lower than in unfavourable rainfall years, due to further dilution of the limited amount of available plant nutrients. On the average, peak herbage production on deep soils is about 2,000 kg per ha. Meat (mainly lamb) production can then average about 55 kg liveweight per ha without the use of supplements. Production per unit area is much lower on shallow upland soils.

1.3. Agro-pastoral systems

The traditional agro-pastoral systems were practiced by both sedentary villagers and nomadic Bedouin in the semi-arid regions of the Middle East for about 5,000 years with little change until the 1950's. In this system, mixed flocks of sheep and goats are maintained for most of the year by grazing common rangeland, non-cultivable hillsides and fallows. In summer the stubble of the grain fields is also grazed and in drought years feed shortages are buffered by grazing the crop itself and by increased migration with the flock. Grazing land, as a rule, is regarded as an uncontrollable natural resource to be exploited as fully as possible while nothing is done to improve its productivity.

The productivity in these semi-arid regions is low, not only because of the low and erratic rainfall, but also because the price of meat, wool and grain has been low compared with the costs of water management, fertilizers and concentrates that are needed to increase the productivity of these systems. However, this is changing rapidly, because extended areas in West Asia and North Africa are now close to large urban centres with considerable affluence and a high demand for animal products, including lamb meat.

Seasonal nutritional deficits in animal nutrition can be alleviated by the use of supplementary feeds. However, this practice is economically feasible only where the local price ratio between lamb live weight and supplementary concentrates considerably exceeds the biological conversion rate. Accordingly, when the price ratio is above 10, it becomes feasible to set the production target close to the genetic potential of the breed as determined by the maximum rates of reproduction and of weight gain. Not only entrepreneurial sheep flocks, but also flocks in the traditional Bedouin and village societies have gradually and steadily moved in this direction. Supplementation tends to increase flock size and to reduce mobility. Consequently, heavy overgrazing occurs in concentrated areas around settlements together with undergrazing in more distant pastures. Generally, no effort is invested to improve pasture production, even though there are a number of feasible technological options.

Natural fertility of many soils in the semi-arid zone is low. Often, average pasture production may be increased at least twofold by improving the fertility of the soils by the use of fertilizers and leguminous species. This increase is even larger in years with normal and above-normal rainfall, but in drought years water remains the constraining factor. Therefore, annual fluctuations in yield are very much larger under improved fertility conditions and it is practically impossible to constantly adapt the size of the flock to this increased variability in primary production. Hence, systems where only the productivity of the land is improved, but no additional concentrates are used, are technically inefficient for lamb production. The only way to utilize the good quality herbage growing in relative abundance in favourable years is to maintain flock size in drought years by feeding concentrates to lambs and ewes.

The present study investigated the possibility that such semi-intensive, integrated crop-livestock systems could be viable in the northern Negev. Subsistence systems were not ignored, but served mainly as a reference and point of departure.

1.4. Perspectives for development

Semi-arid zones are poorly endowed agriculturally in the sense that the production per unit area of land is limited by relatively low and erratic rainfall even in relatively wet years. Yet this is a narrow view of endowment, because an acceptable livelihood does not only depend on the availability of natural resources, but also on the socio-economic context. Given the capital, the necessary land area, and a favourable input-output ratio, agriculture in the semi-arid zone may support a standard of living compatible with the standards of the region. This is especially so in the mediterranean semi-arid zone where proximity to markets and relatively cheap inputs from intensive agricultural by- and waste-products increase the resource base considerably, particularly for integrated agro-pastoral systems. In addition, given the apparently promising future market for small-ruminant products within the region, a feasible agro-pastoral activity will most likely be sustainable, particularly as continuous wheat or barley cropping is falling victim to low international prices and to cumulative phytosanitary problems.

Selection of more intensive agricultural production systems is, in a sense, similar to the selection process in biological evolution: there are many options, but only a select few have survival potential in an environment that imposes constraints in a multi-dimensional context. In addition, major system changes are irreversible, at least in the short run, and once implemented, preclude many other options. Fitness, in relation to agricultural development, means the ability to meet the minimum goal requirements of all the relevant stakeholders in the development process. Thus, considerations of farmers income, return to capital, environmental quality, employment, risk, professional competence, and so on, must all be evaluated concurrently so as to ensure that the selected technologies have a good chance to succeed in a real-world context.

As the range of choice is wide, and the considerations complex, it is hazardous to depend on intuitive or subjective judgement only, even though that is what is usually done, often at great cost. Fortunately, computer power can be harnessed to make the selection of development pathways formally tractable within an interactive dynamic multiple-goal context. While selected pathways may not, in fact, be implemented in the end, at least the options will have been thoroughly analysed and some illusions and pitfalls will have been foreseen and, hopefully, avoided. In addition, the approach can contribute towards more efficient agricultural development planning because it provides a formal communications bridge between the representatives of the biophysical and the socio-economic components of agro-pastoral systems. In that way it allows for an effective interdisciplinary application of scientific knowledge to arid-zone development. That, in short, is what this book is all about.

2. Structure and dynamics of grazing systems on seasonal pastures

I. NOY-MEIR

2.1. Introduction

2.1.1. *Types of plants*

The natural vegetation in the semi-arid regions around the Mediterranean consists mainly of three major plant types or life forms: annual grasses and forbs (therophytes), perennial grasses and forbs (hemicryptophytes), shrubs and trees (phanerophytes). Their relative success and abundance varies between regions and habitats, depending on specific climate and soil conditions (Westoby, 1980). Each has a combination of characteristics which affects its utility for animal production, some positively, some negatively.

Here attention is focussed on regions with a strongly seasonal winter rainfall and a long dry season, like the Negev desert in Israel. Annuals have a life cycle that is well adapted to such a rainfall regime (Fig. 2-1). They also have clear advantages as pasture plants: high shoot growth rates, very high nutritive quality when green and fair quality when dry (Tadmor et al., 1974). They are able to maintain their populations under heavy grazing and large weather fluctuations, due to seeds surviving in the soil. Their main limitation is that they provide green feed for a relatively short period of the year. Often their use of available soil water is incomplete, in particular of moisture stored in deeper layers and off-season rainfalls.

Perennial grasses and forbs are superior to annuals in that they are able to use soil moisture from deeper layers and for longer periods and thus provide green feed for a greater part of the year. However, in semi-arid regions they are more susceptible to drought years when the deeper soil layers are not recharged and the dry periods are long. They are also sensitive to damage by continuous heavy grazing, due to attrition of carbohydrate and nitrogen reserves or of meristems (Breman, 1982). As a result, populations of palatable perennial grasses in semi-arid zones have often been destroyed by temporary overgrazing, particularly in drought years, and have been slow to re-establish. Under grazing mainly those perennial grasses persist which are of relatively low nutritive quality (low protein, high fibre content) and, therefore, are not preferred by grazers (*e.g. Andropogon gayanus* Kunth) or of low availability due to their rhizomatous creeping form (*e.g. Cynodon dactylon* Pers) (Breman, 1982; Noy-

7

Th. Alberda et al. (eds.), Food from Dry Lands, 7–24.
© 1992 *Kluwer Academic Publishers.*

8

green leaf biomass (relative scale)

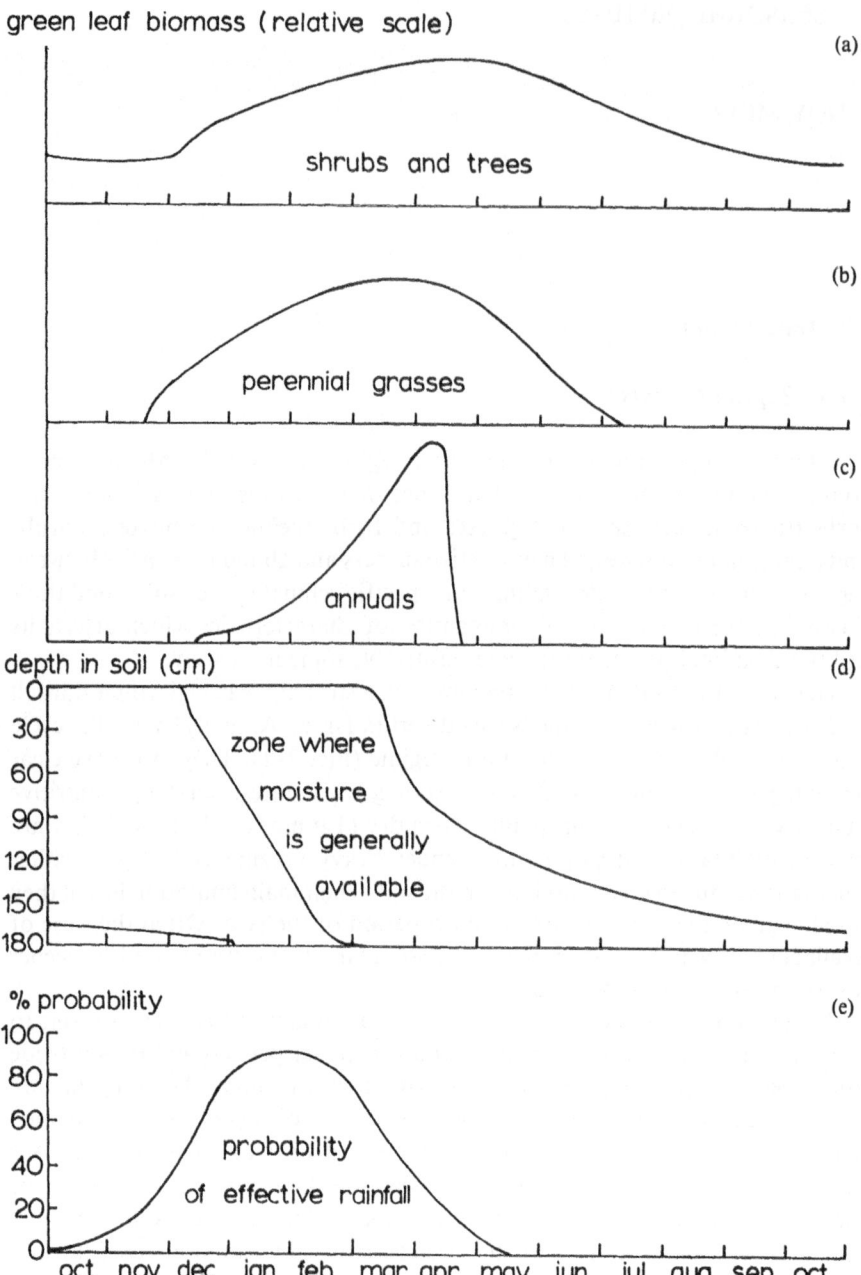

Fig. 2-1. The annual cycle of green biomass of shrubs and trees (a), perennial grasses (b), and annuals (c) in a semi-arid region with strongly seasonal rainfall in relation to the zone in the soil profile where moisture is available (d) and the probability of effective rainfall (e) (diagrammatic, based on data from the Negev).

Meir and Seligman, 1979).

Shrubs and trees in the semi-arid zones usually have root systems which enable them to take up and transpire water continuously throughout the year, and even through drought years, from layers which are either very deep, stony, or saline. This enables them to carry green foliage continuously, even though its amount may fluctuate. In dry periods this foliage is of much higher nutritive value (in particular protein content) to animals that can use it than the dry remnants of herbaceous plants. But intake and digestion of tree and shrub leaves is often limited by high concentrations of secondary substances (tannins, alkaloids) or minerals (salt). The woody tissues of trees and shrubs contain a substantial reserve of carbohydrates which enables them to survive fairly drastic defoliation. However, continuous overgrazing can kill shrubs; if grazing is too conservative much of the primary production is allocated to woody tissue, which is almost useless as feed. The individual plants are long-lived but once mortality has occurred, due to extreme drought or overgrazing, population regeneration from seedlings is slow and may be completely blocked if the seedlings are browsed.

On balance, in regions with a sharply seasonal rainfall the annual vegetation is the most reliable main base of animal nutrition and production, though it leaves severe nutritional problems in the dry season. Suitable species of shrubs and perennials are desirable as complementary components of the vegetation which may alleviate these deficiencies. There is a problem of how to manage grazing on them in a way that will maximize their use without endangering their persistence. In regions where the rainy season is longer or more diffuse, the persistence of perennial grasses and their contribution to animal nutrition will be greater.

2.1.2. *Questions to be dealt with*

This chapter treats the following questions:
a. What attributes of the pasture, apart from total primary production, are relevant for animal nutrition and production?
b. How does grazing during the growing season affect pasture production and biomass dynamics?
c. How is the initial seedling density and biomass determined by plant population processes and how is it affected by grazing?

Other chapters deal with models of plant production in semi-arid regions as determined by weather, soil moisture and nutrients. They explicitly or implicitly assume that:
– the plants are barley, wheat or annual grasses;
– the final yield of shoot dry matter or seeds is mainly of interest;
– the sward is not grazed during the growing season;
– the initial seedling density or biomass is fixed.

Application of plant production models to problems of production and management in grazing systems, or mixed crop grazing systems, requires relaxation of all or some of these assumptions about the vegetation.

2.1.3. *Plant attributes relevant to animal production*

The total annual primary production of the pasture sets an upper limit to its potential contribution to animal nutrition and production. The actual contribution depends on the amount and composition of the herbage consumed every day during the year, which in turn depends on many attributes of the pasture. The pasture attributes which influence animal intake can be broadly categorized into 'availability' and 'quality'. Each of these varies from day to day due to environmental conditions, the development of the plant and the effects of grazing.

2.1.3.1. *Pasture availability*

Pasture availability includes the physical attributes of the vegetation which determine the ability of the animal to find plant material and grasp it with its mouth and the rate at which it can do this. These attributes are:
- the number of plants per unit area;
- their height;
- the amount of leaf and stem material per plant, per area and per volume;
- the vertical and horizontal distribution of this material;
- the amount and distribution of plant structures which impede intake (woody stems, spines).

Ideally, pasture availability should be defined and measured by a parameter such as 'the maximum rate at which the animal can ingest herbage while actually grazing'. Knowledge of this parameter, together with two or three measures of pasture quality and of animal metabolic requirements, can be sufficient for accurate and general predictions of actual intake over a wide range of conditions. However, in the field it is not simple to measure ingestion rate or related parameters, like biting rate and mean weight of bite. In practice, pasture availability is usually measured simply by mean plant biomass per unit area. This may be a reasonable approximation because it depends on plant number, size, height and foliage density (weight per volume).

Ingestion rate also depends on all these attributes and is expected to be closely correlated with biomass for pastures with comparable vertical and horizontal biomass distribution. Empirically, actual intake data when plotted against biomass in a particular kind of pasture usually show reasonable fit to curves which increase at low biomass and then level off. Given this empirical intake-biomass relation, biomass is a good estimate of availability for most practical purposes (Black and Kenney, 1984). However, this relation may differ substantially for pastures with different species composition, different horizontal patchiness, or different vertical structure due to grazing history (Ungar, 1984; Ungar and Noy-Meir, 1988).

2.1.3.2. *Pasture quality*

Pasture quality includes the chemical and physical attributes of the plant material in the pasture, which determine both the ability of the animal to digest it and to use it for its metabolism, and the rate at which it can do this.

A good measure of the nutritional quality of plant material could be the maximum rate at which the animal can extract net metabolizable energy from it, when it is short of energy and feed availability is unlimited. A second measure could be the analogous one for metabolizable protein, and so for any other required nutrient. The rate at which a nutrient can be extracted from a given plant material depends on the nutrient content in the material and on the rate at which the material itself can be digested and assimilated. The crude content of any nutrient is easily measured in vitro, but the estimation of the digestible and net metabolizable fractions requires more elaborate and less reliable measurements.

The rate of digestion, or rate of passage of feed through the digestive system, is also not easy to measure empirically. It is also not theoretically simple, since the processes in different sections of the ruminant digestive tract occur in part simultaneously. The physical rate of exit of material from the rumen is often thought to be the limiting rate.

In any case, the rate of digestion has been found to be positively correlated with the cell content/cell wall ratio, protein or nitrogen content, food particle size, digestibility, and inversely with fibre content. Similar correlations have been found for intake in conditions of unlimited availability (*ad libitum*), which is presumably determined by the digestion rate (except for food of very high quality). Some quality attributes, like protein and fibre contents, affect both nutrient content and digestion rate in the same direction and hence have a doubly strong effect on the nutrient extraction (or retention) rate of plant materials.

For practical purposes percentage digestibility is often a reasonable indicator of digestion rate and of nutritional quality.

2.1.3.3. *Availability, quality and selection*

As a good approximation, the roles of pasture availability and quality in animal nutrition can be treated as a 'law of the minimum' or 'limiting factor' situation. The flow of nutrients in the animal will usually be limited by the slowest of the three consecutive processes:
- ingestion, which depends mainly on feed availability;
- digestion, which depends mainly on feed quality;
- metabolic assimilation.

Thus intake will be determined by ingestion when availability is low, by digestion when quality is low, and by metabolic requirements when both availability and quality are high, or when feed composition is imbalanced.

However, pastures are heterogeneous, consisting of plant parts with a range of levels of availability and quality. Animals can and do select parts of higher availability and/or quality than the average of the pasture, thereby increasing

12

relative "availability" or "quality"

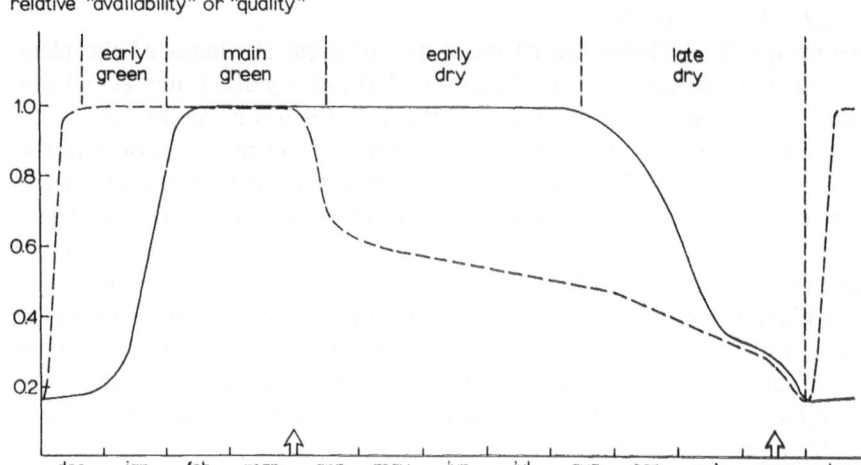

Fig. 2-2. The annual cycle of pasture availability (potential ingestion rate relative to maximum, solid line) and quality (potential digestion rate relative to maximum, broken line) in a winter rainfall semi-arid region and the distinction of four pasture seasons (diagrammatic, based on data from the Negev).

their ingestion and/or digestion rates. Thus intake depends not only on the mean of pasture attributes but also on their distribution (Arnold, 1981; Black and Kenney, 1984); greater variance will allow selection for higher intake (Ungar, 1984; Ungar and Noy-Meir, 1988). Often availability and quality are negatively correlated in the pasture. For instance, shortly after germination the pasture consists of large parts of old straw of poor quality, and small highly nutritious green seedlings. Depending on the selective behaviour of the animal, either quality (digestion) or availability (ingestion) may be limiting intake. The optimal behaviour which maximizes nutrient intake is at the boundary where both limits meet.

2.1.3.4. *The annual pasture cycle of animal nutrition on pasture*
In a temperate, semi-arid winter-rain climate the annual cycle of the pasture can be divided into four phases with respect to animal nutrition (Fig. 2–2; Noy-Meir, 1975b).
a. *The early green season* begins with germination. The new green pasture is of optimal quality (75–80 percent digestibility, 15–25 percent protein), but its ingestion is limited by low availability. As green biomass accumulates, ingestion rapidly increases until the maximum intake rate (limited by metabolic requirements) is reached.
b. *The main green season.* Availability, quality and intake rate are high and all nutritional requirements are fully met. Quality of vegetative parts begins to decrease during flowering and seed filling; thereafter quality drops rapidly as the pasture dries up, shedding the seeds and part of the leaves.

c. *The early dry season.* Dry material is abundantly available but its quality limits digestion rate and intake. Animals can still select a diet of fair quality from the dry material (leaves, fine stems, seeds) (de Ridder *et al.*, 1986).

d. *The late dry season.* As a result of continuous selective grazing and weathering, little fine dry material is left and animals must take in more coarse material (straw) of lower quality and digestion rate. The availability of that material also gradually decreases. The first rains cause a further rapid decrease in both availability and quality.

The potential contribution of an area of pasture to animal nutrition depends on the lengths of these four periods. In particular, the length of the main green season (full nutrition from pasture) is an important and well-defined parameter. Within each of the three other periods there are trends in pasture availability and quality.

2.2. Within-season plant biomass dynamics under grazing

2.2.1. *Leaf area dynamics*

In a grazed pasture, plant production, availability and quality depend not only on the weather and the supply of moisture and nutrients, but also on the actions of the animals: grazing, trampling, excretion. Estimates of plant and animal production from pasture, based on water and nutrient constraints and efficiencies of use measured in undisturbed vegetation, may, therefore, be too high or too low. To understand and evaluate the effect of the grazing animal on production from an area, it is necessary to consider the ecological processes in plant-animal interactions. These are numerous and complex; yet it is possible to identify a few key processes which determine the major effects of practical importance. For instance, defoliation by grazers during the growing season can have a major effect on the rates of plant growth and resource use in the same season.

The direct and immediate effect of grazing in the growing season is a reduction of leaf area and mass and thus of photosynthesis and production rates.

First it is necessary to present the theoretical definitions of variables used in the figures and in the further text. The quantitative relations involved can be discussed with reference to the following quantities (Noy-Meir 1975a, 1978):

V = the amount of green biomass per unit area of land (g per m^2 or kg per ha);

G = the daily rate of absolute net accumulation (growth) of green biomass (g per m^2 per day or kg per ha per day); in the absence of grazing $dV/dt = G$;

g = the maximum daily relative growth rate of green biomass (kg per kg per day or day^{-1}), attained when the amount of green biomass accumulated is still small and shading between leaves in the canopy is negligible; the initial slope $dG/dV = g$.

In the first part of the growing season the growth of the new leaf depends mainly on photosynthesis. The absolute growth rate G increases linearly with V, that is the relative growth rate g is at its maximum. Thus there is a positive feedback loop, which in undisturbed vegetation causes accelerated (more or less exponential) accumulation of leaf area and green biomass.

Continuous defoliation by grazing can slow down this accumulation, arrest it, or even reverse it, depending on defoliation intensity (Noy-Meir, 1975a). There are several plant mechanisms which may compensate for leaf removal by grazing (McNaughton, 1979). The photosynthesis of the remaining leaves increases due to better exposure to light, and sometimes increased photo-synthetic efficiency. Growth of new, efficient leaves is stimulated, often drawing on reserves stored in stems and roots. In this respect there is a difference between an annual pasture and a barley or wheat crop in that in the latter the positive effects of grazing on the formation of new leaves through an increase in the rate of tiller formation are less pronounced. This might at least partly explain the fact that heavy grazing has a far more detrimental effect on small grains than on pasture (see 2.2.4.1).

To analyse the balance between growth and defoliation rates, the following quantities may be defined (Noy-Meir, 1975a):

H = stocking rate or density of grazing animals (animals per ha);

c = daily rate of consumption (intake) of green biomass by a single animal (kg per animal per day);

C = cH, daily rate of consumption by the grazing animals per unit area (kg per ha per day).

When the animal density H and, therefore, also the mean frequency and intensity at which individual plants are defoliated by grazers are moderate, leaf growth between defoliations can effectively compensate. The mean leaf growth rate of the sward (G) will be much greater than the rate of leaf removal (C), and the leaf will continue to accumulate under grazing throughout the growing season. The cumulative leaf production in the grazed sward will not be much less than in an ungrazed sward (Noy-Meir, 1975a, 1978). It may even be greater if in the latter the sward becomes too dense or senescent in the later part of the growing season.

A very different dynamic situation develops under heavier stocking, when the frequency and intensity of defoliation are so high that the rate of regrowth can hardly keep up, and the compensatory mechanisms are not effective. Sward photosynthesis decreases to the point that reserve mobilization can no longer make up for it. Leaf area will accumulate at a much slower than normal rate, or it may stagnate or even deteriorate, which will further reduce the growth rate. Such a 'leaf area crisis' or 'pasture crash' is likely to occur under continuous or frequent defoliation, particularly in an initially sparse sward with low leaf area.

In annual pastures, initial leaf area and biomass are always low in the first few weeks after germination. This is the critical period in which a leaf area crisis will develop if the pasture is grazed too heavily too soon. In such a pasture leaf area, biomass and production will remain low for a great part of the growing season, much lower than in an ungrazed or moderately grazed pasture. Even if the pasture eventually emerges from the crisis and leaf area builds up, it will usually not catch up before the end of the growing season. Cumulative plant production in such a pasture will be greatly reduced (Johnson and Parsons, 1985; Noy-Meir, 1978).

2.2.2. *Deferred grazing*

There are two ways to avoid a leaf area crisis early in the season and maintain a high level of pasture production and availability throughout the growing season. One is to reduce animal density to a level which ensures that the defoliation rate is much less than the growth rate, even in the early part of the season. The other is to defer grazing on the pasture after germination until leaf area and growth rate build up, and then introduce grazers at a correspondingly higher density, always trying to avoid the defoliation rate becoming higher than the growth rate.

When the amount of aboveground biomass is low, the growth rate G is approximately linearly related to green biomass V, $G = gV$, where g is the relative growth rate (which in a pasture well provided with water and nutrients lies around 0.07 per day). The defoliation rate C is the product of stocking density H and the consumption rate per animal c; $C = cH$. The latter is a saturation function of green biomass, which can be approximated by a ramp function:

$$c = \min (sV, c_s)$$

where s is the initial slope of the function ('grazing efficiency')
and c_s is the satiated consumption rate.
To maintain defoliation rate below growth rate,

$$C = cH = \min (sV, c_s{}^*H < gV = G$$

that is either $H < g/s$ or

$$c_s/g^*H < V$$

The question of how long to defer can be approached by modelling the balance of growth and grazing processes (Noy-Meir, 1975b; Ungar, 1984, 1990).

This very simple model suggests the following rules for deferment:
a. if the stocking rate is less than g/s, growth will exceed grazing, even at very low biomass and deferment is not necessary to maintain a positive biomass balance;
b. if the stocking rate is greater than g/s, grazing should be deferred at least until the biomass threshold has been reached at which growth exceeds grazing; this threshold is linearly related to stocking rate and corresponds to a biomass allowance of c_s/g per animal.
Thus the decisions on stocking rate and length of deferment are interdependent.

However, if grazing starts just at the critical threshold thus calculated, the net rate of green biomass accumulation will be close to zero for the rest of the growing season. To ensure continuing and substantial accumulation of biomass under grazing, a higher deferment threshold must be used. How much higher? In the management of experimental grazing systems at Migda in the northern Negev, the critical biomass threshold was multiplied by a 'safety factor' between 1.5 and 2. This rule-of-thumb proved to be generally adequate.

A more rational approach was developed by Ungar (1984, 1990): What is the optimal length (or biomass) of deferment which maximizes the contribution of the pasture to herd nutrition, either in the green season alone, or over the entire year? A model with more realistic growth and consumption functions was formulated and tested in systematic simulation experiments. The results showed that the optimal entry biomass is more or less directly proportional to the stocking rate to be used (see Fig. 6–1). However, it is also markedly affected by pasture attributes and optimization criteria.

If deferred grazing is practiced, the herd must be fully fed outside the pasture in the deferment period. Thus the length of the deferment period determines the amount and cost of supplementary feed in this period. It depends not only on the stocking rate, but also on pasture attributes, in particular the initial seedling biomass V_0, and the relative growth rate g. The higher g and V_0, the shorter the optimal deferment period (Ungar, 1984, 1990).

2.2.3. *Interactions with water, nitrogen and other factors (see also Chapter 3)*

While leaf area dynamics is a central process in pasture production, there are some other important processes which interact with it. The lag in accumulation of leaf area and cover, due to grazing, reduces the transpiration rate of the sward, but increases evaporation from the soil surface. Since evaporation very soon after rain becomes limited by the conductivity of the dry surface layer, the net effect in most situations is likely to be a reduction in evapotranspiration by grazing. The greater amount of soil moisture remaining stored in the soil may allow the grazed pasture an extra period of growth at the end of the season. This will be the case if the primary production of the grazed pasture is limited by soil moisture, as that of the ungrazed pasture. This is another reason why grazing, up to a point, often does not reduce pasture productivity but improves pasture quality and lengthens the green season. However, when the grazing pressure is too heavy, the pasture is not able to use all the soil moisture before pasture growth is terminated by hot dry weather, or by the phenological development of the plants. Heavy defoliation also reduces root growth and the plants are not able to use the deeper soil layers, where moisture is available into the dry season.

These interactions between grazing, leaf area and plant-soil-water processes are easily and realistically represented by the ARID CROP model (Van Keulen, 1975; Van Keulen *et al.*, 1981), with the addition of a simple grazing or defoliation function. Experimental runs have shown that ARID CROP can simulate observed growth curves of grazed pastures, provided that the grazing function used makes allowance for selective grazing of denser patches early in the season (Van Keulen, unpublished).

Grazing interacts also with nutrient processes by removing nutrients from the vegetation and returning part of them in excretions to the soil surface. In the context of semi-arid annual pastures grazed by sheep, faster nitrogen recycling by grazing is important in the annual nitrogen balance (Harpaz, 1975; Noy-Meir and Harpaz, 1977), but it is not clear how effective it is in increasing plant production within a growing season. Removal of nitrogen by defoliation can reduce productivity, in particular in annual plants in a nitrogen-poor soil, which take up most of the available soil nitrogen early in the season, building up high concentrations in leaves, which are diluted during subsequent growth (Seligman et al., 1976; Breman, 1982).

This aspect can be fairly precisely simulated by the PAPRAN model (Seligman and van Keulen, 1981). Grazing also affects the morphology and phenology of plants in ways which may have an important influence on productivity and biomass dynamics (Crawley, 1983). Selective grazing for leaf directly reduces the leaf/stem ratio, but this often is compensated, and probably over-compensated for, by the fact that regrowth after grazing is usually mostly leaf. Similarly, the shoot/root ratio is initially reduced by grazing, but during regrowth most photosynthates are allocated to the shoots and even root-to-shoot translocation may occur. Grazing usually stimulates vegetative growth and reduces flowering and seed filling. Grazing reduces the height and modifies the structure of the sward both directly, by defoliation from the top, and indirectly, by inducing increased tillering or branching close to the ground. These pheno-morphological effects have so far been modelled crudely or not at all, mainly due to scarcity of good empirical knowledge. It is difficult to assess at this stage if, or in which conditions, they have substantial effects on pasture biomass dynamics and productivity.

2.2.4. Growth under grazing in semi-arid annual grasslands in Israel

Biomass growth curves of grazed plots and ungrazed controls were measured at the Migda Experiment Farm in the period 1974–1980 in experiments designed to study specific processes or test specific hypotheses. The pastures used in these experiments were natural annual pastures, the dominating species being *Phalaris minor* Retz and *Hordeum murinum* L. The arable crop used was wheat; sheep stocking densities varied from 10 to 40 sheep per ha. For more details see Benjamin et al. (1978).

From these experiments the following conclusions can be drawn:
- In a more or less optimal situation, with sufficient water and fertilization with nitrogen, potassium and phosphorus, the growth of the young seedlings in the ungrazed pasture is exponential at a rate of 0.07 per day until a green biomass of $V_x = 2,000$ kg per ha is attained; thereafter the growth is more

or less linear at a rate of 140 kg per ha per day. This continues until the end of the green season when weights of around 10 tons of harvestable dry matter per ha have been reached (see also 3.4.1.1 and Figures 3–7a and 3–8a).

- When no nitrogen fertilizer is applied the exponential growth rate is approximately 0.06 per day and $V_x = 800$ kg per ha and the maximum yield lies around 3 tons per ha.
- With a sufficiently long deferment period a stock density of up to 20 sheep per ha does not have much influence on total dry matter production, *i.e.* when the weight of the herbage taken away by the sheep is added to the final harvest of the grazed plots.
- When the stock is taken from the field, the rate of regrowth is the same as that of an ungrazed pasture at the same biomass.
- The growth rate of a pasture grazed with a given number of sheep increases with the length of the deferment period.

From the grazing experiments on a wheat field the following results were obtained:

- The exponential growth rate of an ungrazed field is the same as that of an ungrazed pasture, both with nitrogen fertilization. At a V_x of 2,000 kg per ha there begins a stage of linear growth which continues until a maximum biomass per ha is reached. This linear growth rate in an N-fertilized wheat field is not consistently different from that in an N-fertilized natural pasture in the same year. With sufficient rain the maximum biomass is determined by senescence, which for wheat begins later than with a natural pasture and thus results in a higher maximum biomass; under water shortage the maximum biomass is about the same.
- Grazing reduces total dry matter production (final harvest plus amount grazed away) in wheat more than in pasture. The possible reasons are:
  a. greater ability of sheep to select leaves from stem in wheat compared to natural grasses;
  b. lesser ability of wheat to compensate the grazed leaf area with an increase in the rate of tiller formation and a more prostrate growth form (see Section 2.1.1).

2.2.4.1. *Effects of grazing on primary production*

In an annual crop or pasture growing without disturbance, the cumulative primary production over a growing season can be fairly closely estimated by the maximum yield which is attained at the end of that season. In a sward which is harvested or grazed during the growing season, the amounts of dry matter removed by harvesting or grazing must be added to the final yield to obtain a comparable estimate of annual primary production. The difference between the latter and the production of an ungrazed field is a measure of the effect of the grazing or cutting treatment on plant production, which may be stimulation, damage, or no effect.

The Migda experiments provide data on this effect in annual swards of different compositions, under various grazing treatments, in several years with different rainfall distributions (Table 2–1).

Table 2–1.
*Estimates of reduction of primary production (standing biomass + consumption) by grazing in the
growing season in annual grassland and crops at Migda.*

Year	field	grazing pressure (ewe-days per ha)	% reduction
1974/75	13 NPK2	1,200	0
	11-wheat-1	800	30
	11-wheat-2	1,300	50
1975/76	13 NPK	600-800	30
	13 PK	200-400	0
1977/78	1,4,11,41,12	200-300	0
	13 NPK	600	30
	13 PK	600	20
1978/79	1,4,11,41,12	300-500	0
1979/80	1	400	30
	4,11,41	500-800	20
	12	1,000	30
	13 NPK1-early	65	20
	13 NPK2-late	42	30
	13 PK1-early	55	0
	13 PK2-late	65	30
	151-154 barley	80-100	40

In general, grazing either had no significant effect on plant production or reduced it by about 20 to 40 percent. A stimulation of about 10 percent was measured in some cases, but it was not statistically significant.

Natural annual pastures, with and without nitrogen fertilizer, were not affected at all by grazing (with deferment) in some years, while in other years reduction in production by 20 to 30 percent was found. There is no simple correlation with the yearly rainfall, but it appears that the difference is that in some years the grazed pasture is able to utilize late rains, while in other years it is not able to do so, perhaps depending on the amount of live roots remaining at the time of the rain.

In years when grazing did reduce pasture production, the magnitude of the effect was not clearly related to stocking rate over a wide range of 3.3 to 15 ewes per ha or 30 to 100 ewe-days per ha in the growing season. This is perhaps less surprising if one considers that the beginning of grazing was deferred to a green biomass threshold proportional to the stocking rate. It confirms that the general principle and the particular formula used were by and large effective in preventing early season damage to pasture growth. However, in 1979/80 the reduction appeared greater in two treatments: light continuous grazing without

20

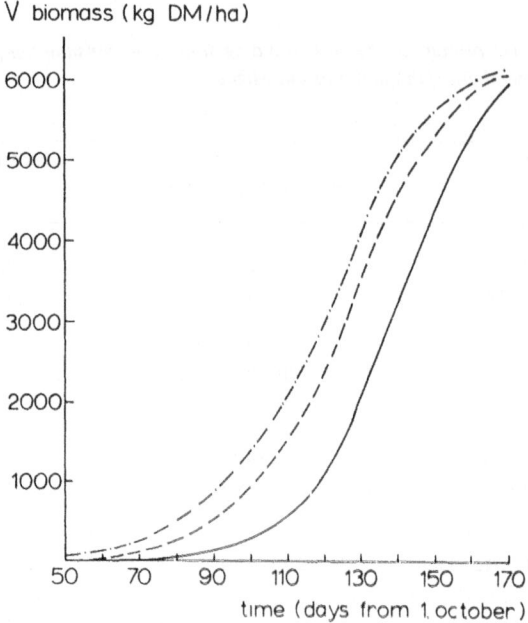

Fig. 2-3. The effect of initial biomass (V_0) on biomass accumulation in an ungrazed and unfertilized annual grassland (Migda, 1972/73). Simulated results with ARID CROP (van Keulen et al., 1981);
-- $V_0 = 10$ kg ha^{-1}; - - - $V_0 = 50$ kg ha^{-1}; - · - · - $V_0 = 100$ kg ha^{-1}.

deferment (3.3 ewes per ha) and very heavy grazing with deferment (15 ewes per ha). In this year, the third year of fixed grazing treatments in the experimental fields, there also was a tendency for pasture production in ungrazed exclosures to decrease in proportion to the stocking rate to which the field was exposed in the two previous years. This may reflect slow effects of heavy grazing on plant populations or soil surface conditions (see below).

The primary production of wheat and barley is reduced by heavy grazing in the main growing season, even after deferment, by 30 to 50 percent; considerably more than natural pasture in the same conditions. This may be due to the growth habit of the cereals, which allows sheep to pick leaves off the stem more effectively. However, a short grazing period on wheat or barley up to 45 days after emergence has little or no effect on the final yields of dry matter and grain (see Section 4.2.3).

2.3. **Between-season plant population dynamics and the effects of grazing**

2.3.1. *Initial seedling biomass*

The initial green biomass at the beginning of a growing season (V_0) can be measured when seedling emergence is complete. It depends on the number and weight of seeds available in the soil and on the proportion of them that have actually germinated and successfully emerged. Since in the first part of the season plant growth rate depends strongly on leaf area and green biomass, accumulation is approximately exponential; the initial biomass determines the biomass at any time after germination at least for a month or two. After that, other factors, like soil moisture shortage or light limitation at high leaf area, may compensate for and reduce the advantage of a high-V_0 pasture over a low-V_0 one in terms of final yield and seasonal primary production (Fig. 2–3). Thus final dry matter and seed yield of an ungrazed sward will not be 'limited' and not be sensitive to initial biomass over a wide range of the latter; only when seedling biomass is very low will the final yield be substantially affected. This is the biological basis of the rather low sowing densities that can be used in crops like wheat and fodder legumes.

However, in a pasture to be grazed during the growing season, the contribution of a pasture to animal nutrition and production depends not only on the final primary production, but largely on the level of green feed availability during the entire season, and on the rate and duration of grazing that can be maintained by the pasture without availability being reduced, so that it limits intake. Initial biomass, together with the relative growth rate, is a critical factor in the early-season leaf area balance, which determines green pasture intake under both continuous and deferred grazing. Under heavy grazing the final primary production can also be quite sensitive to initial biomass.

In the Migda experiments initial biomass was sometimes measured directly just after emergence; otherwise it was estimated by extrapolating the exponential growth curve measured in the period 3 to 6 weeks after emergence backwards to the presumed date of emergence (5 to 10 days after the rain which caused germination). The indirect estimate may be as good as the direct one, since harvesting of small seedlings is difficult and may often lead to an underestimate.

Compilation of the initial biomass estimates over years and fields (Table 2–2) shows most values in the range 3 to 30 kg per ha. In pasture which had not been cultivated or grazed in the growing season for several years it usually was 20 to 30 kg per ha, but occasionally it was much higher (80 to 100 kg per ha), in particular in the part dominated by *Hordeum murinum* L. In the experimental grazing systems initial biomass was usually about 10 kg per ha or less. This can be attributed to the effects of disking the fields at the beginning of the experiments and then grazing them heavily year after year. Surprisingly, there seems to be no strong relation between initial biomass and rainfall in the current or the previous year. However, after high rainfall and high production in 1979/80 initial biomass in a lightly grazed field (and that alone), suddenly jumped to 70 kg per ha.

The initial biomass depends on the number of seeds germinating and on seedling weight. To identify the factors and processes involved in determining the germinating seed population density, the dynamics of plant and seed

Table 2-2.
Estimates of initial green biomass (kg per ha) in different fields at Migda.

	experimental pasture (13)[*]				grazing trial pastures				
	NPK1	NPK2	PK1	PK2	1	4	11	41	12
1974/75	80	30	20						
1975/76	30	25	25	35					
1977/78	15	15	10	15	32		3		4
1978/79					8		4		4
1979/80	20	15	20	15	16		10	15	7
1980/81					70		4		3

[*] Field 13 had four subdivisions, two that received generous annual dressings of NPK fertilizer (NPK1 and NPK2) and two that received PK only (PK1 and PK2).

populations must be studied around the year. This was done in a three-year study in the experimental grazing systems at Migda (Loria, unpublished).

2.3.2. *Plant population dynamics*

In the course of the growing season the initial biomass of seedlings is multiplied at least one-hundred-fold, often by several hundreds. But of the large amount of biomass that has been produced at the end of the season, only about one percent or less re-emerges as seedlings at the beginning of the next growing season. How does this drastic reduction come about and what influences it? To understand this it is necessary to follow the various stages of the annual cycle of plant populations, and to obtain quantitative estimates of the numbers and biomass of plants and seeds at different stages, and of processes and rates.

A fairly constant 30 to 50 percent of the seeds present in the soil in autumn sucessfully germinated and emerged as seedlings; thus the initial seedling biomass is consistently a similar proportion of seed biomass in the soil.

Survival of seedlings and plants in the vegetative stage varied from almost 100 percent in some populations to 60 percent in very dense populations in unfavourable conditions, *e.g.* dry soil. The standing biomass at the end of the growing season is reduced by sheep grazing by 25 to 50 percent. However, in the worst drought under heavy grazing it was in the order of 1,000 kg per ha, compared with the initial biomass in the order of 10 kg per ha.

The proportion of total biomass found in the seeds was usually in the range 15 to 45 percent for annual grasses and 5 to 15 percent for annual forbs (de Ridder *et al.*, 1981). Thus the average seed ratio of fields in a year was strongly affected by species composition and varied in the range 10 to 40 percent.

As a result of its effects on vegetative biomass, on allocation to seeds and direct seed consumption, very heavy grazing (15 sheep per ha) reduced seed yield per area to 20 to 40 percent of the yield in the same year in ungrazed exclosures. In lightly grazed fields, the seed yield was similar to that in exclosures. However, by far the greatest reduction in plant populations and biomass occurred during the dry summer months (April-October), in both grazed and ungrazed plots. Only 3 to 10 percent of the seeds produced in April are found in October, and most of these are buried in the soil or covered by litter. The disappearance of seeds exposed on the surface is rapid and begins during dispersal; in some years 80 percent of the seeds had disappeared by May and 90 percent by July. Most of them were apparently collected by ants (*Messor* spp). Observations and experiments on ant foraging rates, ant nest density and capacity confirmed that ant foraging can, by itself, account for the observed rates of seed disappearance as well as for the total amount. Since seed survival in summer is the bottleneck in the population cycle, any factor causing variation in it is critical in determining the seed stock available for germination and the initial seedling biomass.

The effects of grazing on seed survival are complex. Survival was generally lower under heavy than under moderate grazing. This was attributed not so much to direct seed intake by sheep as to soil surface compaction by heavy grazing in the wet season, which allows fewer seeds to enter the soil before being collected by ants. On the other hand, under moderate grazing seed survival was as high and often higher than in ungrazed enclosures. This was apparently because trampling by sheep during and just after dispersal helped to bring seeds into the soil, thus escaping ant predation.

Thus the effects of sheep grazing on seed survival and on initial seedling biomass are to a large extent indirect effects, through ant foraging efficiency.

What are the practical implications of these features of plant population dynamics in such grasslands for management of grazing systems? The grassland can tolerate fairly heavy sheep grazing (10–15 ewes per ha) in the main growing season including the period of flowering and fruiting (February-April) and in the early dry season (May-July), without causing a serious net reduction in seedling biomass in the following season. A serious reduction can, however, result from soil compaction by heavy grazing in winter and from reduction of plant litter below 500 kg per ha by excessive grazing in late summer and autumn. Avoidance of these situations can be achieved by extending the period of pasture deferment, which is anyway recommended from considerations of early season biomass dynamics. Disking of the pasture corrects surface compaction but markedly reduces initial biomass in the immediately following season. Thus frequent total disking should be avoided. Cultivation in narrow strips at the time of seed dispersal every few years is preferable; addition of grass seeds on this occasion may prove to be economically advantageous.

2.4. Conclusions

As has been discussed in detail in this chapter, the special attributes and dynamics of pasture vegetation can cause levels of primary production of pastures, and particularly of useful primary production, to deviate widely from those predicted from crop models and abiotic variables. To take full account of the known complex processes and interactions in pasture-grazer systems, it is necessary to construct fairly large models with many parameters to be estimated. This is feasible, or possible, only in systems where most of the processes have been studied experimentally and quantitatively.

However, within certain restrictions, production models developed for annual crops (such as wheat) in relation to water and nitrogen can, in combination with fairly simple additional models or procedures, give predictions of total and useful primary production in grazed pasture that are sufficiently accurate for practical purposes.

The conditions are:

a. the pasture is dominated by annual plants, or by perennial herbaceous plants in which translocation to and from storage organs in the course of the growing season is small compared to assimilation, even under grazing;

b. grazing in the growing season is limited to periods when green plant biomass is sufficient to maintain both herbivore consumption and leaf area accumulation at close to maximal rates;

c. initial biomass at the beginning of the growing season, as well as the distribution of biomass between main availability and quality classes at several times during the season, are measured in the field and need not to be predicted by models.

3. Moisture, nutrient availability and plant production in the semi-arid region

H. VAN KEULEN and N.G. SELIGMAN

3.1. Introduction

Semi-arid regions are characterized by low and erratic rainfall that varies greatly from year to year. Rainfall is either monomodal, concentrated in one season or may be bimodal, distributed between two seasons. Dry periods of variable length and variable intensity occur between the rainy periods. The consequence of this weather pattern is that growth of the vegetation is limited to periods when moisture available from precipitation can satisfy the minimum transpiration requirements of the vegetation.

Total annual rainfall is only one aspect of the water supply of plants and in itself is not a sufficient indicator of water availability and its effect on production. Distribution and intensity of rainfall determine the partitioning of the water between runoff and infiltration on one hand and between soil surface evaporation and crop transpiration on the other. The prevailing climatic conditions during the crop growth period influence the efficiency of water use: higher temperatures and lower humidity lead to greater transpiration losses at identical assimilation rates. Consequently, when water availability limits growth, summer rainfall is used at lower efficiencies than winter rainfall for similar photosynthetic systems. Soil physical properties determine the infiltration capacity and hence the degree of runoff, the rates of transport of water through the soil and the moisture storage capacity. In order to understand the effects of water availability on agricultural production it is necessary to describe quantitatively the water balance in the soil-plant-atmosphere system, the reaction of the plant to variable periods of water deficiency and its consequences for crop and pasture production.

Even though water deficiency is the dominant characteristic of arid and semi-arid regions, there are situations where growth is restricted by nutrient availability, especially where rainfall distribution is strongly seasonal. Deficiency of any plant nutrient can limit growth, but nitrogen and phosphorus deficiency are the most common. In the present study only nitrogen availability will be discussed as the dominant nutrient determinant of growth in most of the semi-arid lands around the eastern Mediterranean Basin. The relationships between water, nitrogen and plant growth are formalized in dynamic simulation models that permit comprehensive analysis of different situations.

25

Th. Alberda et al. (eds.), Food from Dry Lands, 25–81.
© 1992 *Kluwer Academic Publishers.*

3.2. The water balance

3.2.1. *Rainfall*

Rainfall intensity in semi-arid regions can be very high and variable. For accurate estimates of runoff and infiltration, continuous recording of rainfall intensity is necessary. As such detail is seldom available, analysis of rainfall regimes is usually based on seasonal totals, monthly totals or at best on daily values. Many attempts have been made to establish relations between rainfall and production, both for natural vegetation and for arable crops. As a rule, such analyses do not probe the underlying mechanisms, but are based on empirical regression equations giving statistical relations between rainfall characteristics and yield. The simplest approach is to relate total seasonal rainfall to yields measured at a given site or region (Le Houérou *et al.*, 1988; Le Houérou and Hoste, 1977; Lof, 1976; Breman, 1975; Lomas and Shashoua, 1973; Lomas, 1972). Somewhat more sophisticated multiple-regression models relate yield to rainfall during different periods of the growing cycle (Zaban, 1981; Lomas and Shashoua, 1974; Baier and Robertson, 1967). Zaban (1981) has developed a regression model between yield and monthly evapotranspiration estimated by a simple model that also considers class A pan evaporation. Such methods are often fairly reliable within the region where they have been developed, but are generally unsatisfactory for extrapolation to other regions with different soils, weather and management practices. A more general approach can be based on the use of conceptual models in which an estimate of water availability is based on the physical and physiological principles underlying crop water relations.

A serious problem in the analysis of rainfall regimes in semi-arid regions is the substantial spatial variability of precipitation even over short distances (Stroosnijder and Koné, 1982; Shanan *et al.*, 1967). This implies that for accurate analysis of yield-rainfall relations, precipitation must be recorded at or very close to the site of interest. Even then, rainfall gauges spaced as close as some tens of meters apart may yield statistically significant different values, especially with respect to the temporal distribution of precipitation. However, weather variables, including rainfall, are usually recorded some distance from the site of interest and the inherent deviations between the recorded and actual data are then an unavoidable source of error.

3.2.2. *Interception*

Part of the rain is intercepted by the crop and evaporates directly from the leaf surfaces. The effect of interception depends on rainfall distribution and intensity and on the properties of the vegetation. The amount of water that can effectively be intercepted by the vegetation before it starts to 'drip' is a function of the canopy surface area (Rijtema, 1965). For a given crop size, interception will be greater with many light showers, than with few heavier ones.

The quantitative effect of interception on the soil water balance is difficult to establish. During periods when adhering water evaporates from the leaf surfaces, transpiration is lower than when the leaf surfaces are dry, which partly compensates for reduced infiltration. The degree of reduction in transpiration depends on such factors as crop surface roughness, prevailing wind speed and the apparent diffusion resistance of the crop surface. In most situations, interception must be neglected in water balance studies, because of uncertainty in determining its magnitude, the doubtful accuracy of gauged precipitation data and the effect of reduced transpiration from wet leaves. This neglect may not be serious in crops and grass canopies (Penman, 1963) but is probably important in the case of trees, where 'steered drip' may increase the amount of water available to the plant by concentrating it near the trunk (Dolman, 1987). This subject is, however, outside the scope of the present study.

3.2.3. *Runoff*

Not all the water reaching the soil surface necessarily infiltrates into the soil. Some is temporarily retained in small ponds caused by the irregularity of the soil surface, the surface storage capacity. When the surface storage capacity of the soil is exceeded, water starts to flow from the site where the rain fell, especially when the terrain is sloping. Only a very light slope is necessary to initiate the process. This can cause redistribution of water on a small scale as a result of microtopographic heterogeneity caused by geomorphological factors, human activity like ploughing or animal activity such as the building of ant nests and the construction of underground channels by rodents. In many semi-arid regions runoff occurs on a much larger scale and precipitation is transported far from its original 'impact site'. As a result, actual infiltration at the site of origin can be much smaller than precipitation.

The infiltration capacity of the soil surface is governed by the physical properties of the top soil. These properties are related to the texture of the soil, *i.e.* the particle size distribution, that determines the size of the pores through which the water can flow. In semi-arid regions the properties of the topsoil are, however, not always stable. Under the influence of raindrop impact, a crust with low water permeability can be formed on the soil surface (Morin and Benyamini, 1977). As rainfall proceeds, crust formation continues, the infiltration capacity of the soil decreases and the probability of runoff increases. Loamy soils with a rather narrow particle size distribution are particularly susceptible to crust formation (Hoogmoed and Stroosnijder, 1984; Stroosnijder, 1982). On such soils more than half of the annual precipitation may flow off to low lying areas or natural depressions. Kinetic energy of raindrop impact on the soil surface is dissipated by organic material that can serve as a mulch and crusting can be delayed or even avoided. At the beginning of the rainy season when rainfall intensity can be high, and soils are bare as a result of cultivation or overgrazing, the effect of raindrop impact can be severe and runoff can be considerable, especially on sloping sites. Estimates of the minimum amount of organic material necessary to protect the soil sufficiently, have been made for various situations, but the amount needed can vary with rainfall conditions. As a rule, relatively small amounts of organic material, less than 1,000 kg per ha, have been found sufficient to reduce runoff significantly.

Sometimes infiltration capacity is decreased by a hydrophobic film on the soil surface that is composed of algal waste products and living hyphae (Rietveld, 1978). In general, the hydrophobic properties disappear relatively quickly after rewetting, but a combination of crust formation and algal mat can seriously reduce the infiltration capacity and cause severe runoff losses.

In many regions where rainfall alone would be too low to enable successful arable farming, runoff is a process that is used to establish viable agricultural systems. Such 'water harvesting' systems (Reij *et al.*, 1988; Evenari *et al.*, 1971) are outside the scope of this study.

Computer models that describe runoff on the basis of topography, soil physical properties and the effects of rainfall on these properties have been developed (Rietveld, 1978; Hillel, 1977; Seginer and Morin, 1970). As a rule, the data requirements of such models are very heavy, and include continuous recording of rainfall intensity. In addition, accurate simulation of soil physical processes requires time intervals of the order of minutes or seconds, as the time constants of these processes are very small (de Wit and van Keulen, 1972). These characteristics make these models incompatible with crop growth models that have to cover a whole growing season. For the latter, the use of highly simplified approximations is necessary.

3.2.4. *Infiltration*

Infiltration of moisture into the soil is governed by potential gradients as water moves from wetter to drier soil. The potential gradient consists of two components, the matric forces of the soil, that are dependent on soil moisture content and the force of gravity. When water infiltrates into a relatively dry soil, matric forces are the major driving force; when water infiltrates into a relatively wet soil, the gravity forces gradually take over. The rate of infiltration also depends on the hydraulic conductivity of the soil, or its ability to transport water, which increases with soil moisture content until all the pores are filled with water. The maximum infiltration capacity of a soil is determined by its saturated hydraulic conductivity.

3.2.5. *Redistribution and drainage*

When infiltration into a soil with a very deep groundwater table stops, a gradient in soil moisture content is established with the highest values near the soil surface and the lowest values deeper in the profile. At that moment, redistribution of moisture begins and tends towards an equilibrium situation where the soil moisture distribution in the profile reflects the gradient of gravity. In the field, the situation is more complex because water evaporates from the soil surface concurrently with redistribution, and there is continuous adjustment of the soil moisture potential. In shallow soils, subsurface drainage can occur, even in semi-arid regions. When the soil is deep, only redistribution or internal drainage is of importance and water does not normally drain beyond the potential rooting depth of the vegetation, except in low-lying areas and depressions where runoff water accumulates.

3.2.6. *Soil surface evaporation*

Once water has entered the soil profile, the most important cause of non-productive water loss is direct evaporation from the soil surface. Under arid and semi-arid conditions with moderately deep profiles, where deep drainage is of no importance, water can leave the soil only by evaporation and transpiration. The ratio of direct soil surface evaporation and transpiration is, therefore, of prime importance in determining the overall water use efficiency, *i.e.* the amount of dry matter produced per unit of precipitation or irrigation. A large part of the scatter observed in statistical analysis relating yield and rainfall in different years or in different regions can be explained by the varying portion of evaporation in the total water balance as a result of differences in environmental conditions, rainfall patterns and vegetative cover.

To sustain evaporation from a porous body like a soil, three conditions must be satisfied. Firstly, energy must be supplied to meet the latent heat requirements for the transfer of water from the liquid phase to the vapour phase. Secondly, the vapour must be transported away from the zone of evaporation by diffusion and/or convection. Thirdly, there must be a continuous supply of water to the evaporating site. The first two conditions are influenced by external factors, such as the level of irradiance, wind speed, air temperature, etc. These factors together determine the evaporative demand. The third condition is dependent on the physical properties of the soil that determine the rate at which water can be transported through the profile to the surface where evaporation occurs. The rate of soil surface evaporation is determined either by the evaporative demand when the soil surface is wet, the so-called 'constant-rate stage', or by the ability of the soil to transmit water to the surface, as the surface dries out, the 'falling-rate stage'. When rainfall is distributed over a larger number of light showers, the soil surface is wet for much longer periods of time and evaporation losses are higher than when rainfall is concentrated in fewer heavier showers. In the field, surface evaporation decreases as the vegetation cover increases, because less radiant energy reaches the soil surface. At the same time, the vegetation cover increases the resistance to vapour transport from the soil surface to the atmosphere, and the rate of transport declines. Experimental evidence has shown that the evaporation rate declines more or less exponentially with increasing leaf area index of the vegetation (Goudriaan, 1977; Ritchie, 1972).

3.2.7. *Transpiration*

Water is transpired by plants mainly as a side effect of assimilation that requires open stomata to allow entry of CO_2. Transpiration is for the major part an inevitable loss although it has functions that are useful for a plant, like cooling and transport of nutrient elements from the soil to the aboveground plant parts (Gale and Hagan, 1966). The rate of transpiration from a vegetative cover depends on the leaf area of the vegetation and on the rate of water loss per unit leaf area. The total amount of radiant energy absorbed by the leaf governs the energy balance of the leaf and is equal to the incident radiation intensity minus the part that is reflected by the leaf surface. This absorbed energy is dissipated by evaporative heat loss for transpiration and by sensible heat loss, *i.e.* the thermal radiation emitted by the leaf to its surroundings. Evaporative heat loss is proportional to the difference in vapour pressure between the evaporating site, *i.e.* the saturated vapour pressure at leaf temperature and the vapour pressure in the atmosphere, and inversely proportional to the resistance to vapour transport from the stomatal cavity to the ambient air. This resistance is composed of two components, the stomatal resistance and the boundary layer resistance above the leaf. The first resistance is inversely proportional to the degree of stomatal opening and is dependent on light intensity, water status of the vegetation and, in some cases, the concentration of CO_2 in the stomatal cavity. The boundary layer resistance is a function of the width of the leaf and the prevailing wind velocity.

The rate of sensible heat loss is proportional to the difference in temperature between the leaf and its surroundings and is inversely proportional to the resistance for heat transfer, which includes only the boundary layer resistance. The partitioning between the two components of heat loss can be found by solving the energy balance for the transpiring surface. Detailed treatments of the energy balance, both for individual leaves and crop surfaces, are available in the literature (*c.f.* Goudriaan, 1977; Slatyer, 1967; Penman, 1948). At a lower level of resolution, semi-empirical equations have been developed to calculate crop transpiration from environmental conditions and crop characteristics.

Comparable vegetation canopies transpire more in situations with summer rainfall than in situations with winter rainfall because the rate of water loss through transpiration is proportional to the vapour pressure difference between the vegetation surface and the ambient air. In summer rainfall regions, temperatures during the growing season are generally higher and relative humidity is lower than in winter rainfall regions, so that transpiration is higher even though the values of the resistances to water vapour exchange may be identical.

3.3. The soil nitrogen balance

The quantitative relations between water and nitrogen in an arid environment vary widely from season to season and within the annual growth cycle. Nitrogen availability can be critical to growth during the vegetative growth phase of annual plants, but less so during the reproductive phase (Kanemasu, 1983). Nitrogen availability can change during a growth cycle from abundance to scarcity and the other way around in accordance with the complex nitrogen transformations in the soil. Phosphorus, on the other hand, is less liable to extreme fluctuation and when it is deficient, its influence on plant growth becomes dominant (Penning de Vries and van Keulen, 1982). The ephemeral abundance or scarcity of dynamic growth factors or the more pervasive abundance or scarcity of the more stable factors are eventually translated into a production level that varies from season to season. Drought may prematurely end a growing season but earlier in that season growth could also have been restricted by nitrogen deficiency when water was available. In that case, growth would have been limited by nitrogen availability despite the overriding effect of the drought (van Keulen, 1975).

In any but the more extreme cases of aridity, the water use efficiency (see 3.4.1.2) of available rainfall that infiltrates into the soil is reduced by the lack of nutrients, primarily nitrogen and phosphorus, but also by other elements like boron, molybdenum and sulphur (Jones, 1963, 1964; Russell, 1958; Williams *et al.*, 1956). In the following subchapter the particularly elusive case of nitrogen availability and its significance to plant growth and production in a seasonally arid environment will be discussed.

3.3.1. *Aridity and the nitrogen cycle*

Water and its effect on the soil environment is the prime mover of many nitrogen transformations in the soil. Not only does it provide the necessary milieu for microbial activity but it can physically move nitrogen from one part of the soil to another. Excess water can exclude oxygen from parts of the soil and cause anaerobic conditions that lead to denitrification and loss of nitrogen as elemental gas or nitrous oxide to the air. As it drains into deeper layers, water can carry nitrate ions with it, sometimes beyond the reach of the plant roots. Rainfall can also introduce nitrogen into the soil. When the amount of water entering the soil is reduced, as it is in arid environments, its impact on nitrogen transformations is also reduced. Less water is available for drainage and loss to deeper soil layers. Soil is wetted to saturation for relatively short periods during the season so that anaerobic conditions are rare and denitrification is potentially less than in more humid conditions (Fillery, 1983; Bartholomew and Clark, 1965). When no water enters the soil for an extended period and ammonia ions accumulate in the top soil layers, loss by volatilization can be significant, especially when the pH is high, as is the case in many arid soils (Freney *et al.*, 1983). In the deeper soil layers ammonia nitrogen is less liable to loss by volatilization and drainage because it is adsorbed on the clay complex. Sometimes ammonia can be lost directly from the vegetation (Farquhar *et al.*, 1983; Lapins and Watson, 1970).

Deep arid, non-alkaline soils are, in a sense, a nitrogen trap. Losses from denitrification, leaching and volatilization are severely restricted so that most of the nitrogen that enters the soil can leave it only via uptake by the plant. Consequently, even though nitrogen can limit growth in arid conditions, the efficiency of nitrogen use for plant growth may be high. This possibility will be examined in the light of experimental evidence and various implications will be studied with an appropriate simulation model.

3.3.2. *Soil-nitrogen transformations*

The major transformations that control the nitrogen balance in the soil are those that convert the nitrogen from organic to mineral, from mineral to gaseous, and vice versa.

a. Organic → mineral. This transformation is termed mineralization and is mediated mainly by microbial activity which, in turn, is dependent on soil environmental conditions as well as on the availability and chemical composition of the organic substrate. Of particular importance is the carbon/nitrogen ratio of the substrate.

b. Mineral → organic. This is termed immobilization and can be divided into two stages: utilization of the mineral N by the micro-organisms and higher plants for growth; and eventual decomposition of the organic residues and partial incorporation into the stable organic soil N fraction.

c. Mineral → gaseous. This transformation is by microbial denitrification of nitrate (NO_3-N) to gaseous N_2 and N_2O under anaerobic conditions and by volatilization of ammonium ions to ammonia under conditions of high pH.

d. Gaseous → mineral. This transformation is the initial stage of biological nitrogen fixation by micro-organisms. The reduced N is incorporated into amino acids and eventually can become part of the available N in the soil via various pathways. Physico-chemical reduction of gaseous N into ammonia occurs in the atmosphere during electrical storms, and can then enter the soil with precipitation.

3.3.3. *The soil nitrogen balance at equilibrium*

The terms of the N balance in a semi-arid environment have been analysed by Noy-Meir and Harpaz (1977) by means of a highly simplified model (Fig. 3–1). The assumption of a 'soil N trap' allows simplification of the model which consequently ignores gaseous loss or loss by leaching beyond the rooting zone. The model defines a set of conservation, flow and equilibrium equations that are used to calculate the equilibrium levels of the main N pools under a set of different management systems. These include an unmanaged grassland 'grazed by wildlife', cropping systems with different levels of biomass utilization with

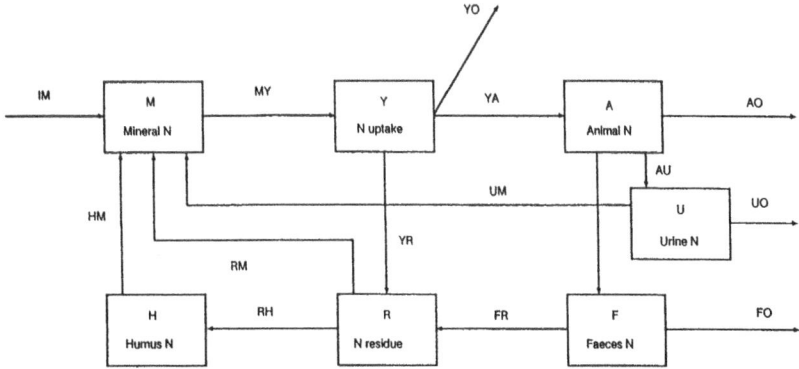

Fig. 3-1. Schematic representation of a simple model of nitrogen dynamics in a semi-arid ecosystem.

Rectangles represent nitrogen pools
Arrows represent nitrogen flows
IM   inputs of mineral nitrogen (rain, dust, N-fixation, fertilizer)
MY   uptake by plants
YO   removed by harvest
YR   from plants to organic residues
RM   mineral N released from decomposing organic residues
RH   from organic residues to stable organic material
HM   mineralized from stable organic material
YA   consumed by grazing animals
AO   removed in animal products
UM   from urine to soil mineral N
AF,
AU   excreted by animals as faeces, urine
FO,
UO   leaving the system from faeces, urine
Note: In order to set up the conservation, flow and equilibrium equations, it is necessary to assign values to the following parameters:
u the fraction of N in the vegetation that is harvested; $(1-u)$ is the fraction that is transferred to the organic residues $(u = 0.2, 0.4, 0.6)$
m the fraction of N in the organic residues that is mineralized rapidly; $(1-m)$ is incorporated into the stable soil organic N fraction $(m = 0.6)$
f fraction of total available mineral N taken up by the vegetation during the year $(f = 0.9)$
d fraction of stable organic N mineralized each year $(d = 0.002)$
x fraction of consumed N excreted as faeces $(x = 0.27)$
y fraction of consumed N excreted as urine $(y = 0.63)$
v fraction of faeces N lost from the system $(v = 0.6)$
w fraction of urine N lost from the system $(w = 0.6)$.

and without fertilizer use, and a managed livestock/pasture system. The pool values and the production of total plant biomass, grain, straw and animal product were found to be quite realistic and appear to support the main assumptions of the model (Table 3–1). So, for example, calculated grain yields of 0.44 to 0.67 ton per ha in the cultivated systems with no addition of fertilizer

34

Table 3-1.
Nitrogen pools and productivity of simple agro-pastoral ecosystems in a semi-arid climate at different levels of utilization intensity (u).

System no.	1	2	3	4	5	6
Fertilizer (kg N ha^{-1})	0	0	0	90	90	0
Utilization intensity (u)	0.2	0.4	0.6	0.4	0.6	0.4
	(1)	(1)	(1)	(2)	(2)	(1)

Nitrogen pools (kg ha^{-1})

	1	2	3	4	5	6
M	5.6	2.7	1.8	19	15	4
H	8,000	3,000	1,350	3,000	1,500	5,000
dH	0	0	0	+34	+19	0
Biomass production (kg DM per ha, 1 % N)	5,000	2,500	1,667	16,600	13,500	3,900
Harvested grain (kg DM per ha, 1.5 % N)	0	667	444	4,400	3,550	0
Harvested straw (kg DM per ha, 0.75 % N)	0	0	444	0	3,550	1,560(3)
Animal production (kg liveweight per ha, 3 % N)	0	0	0	0	0	53

(1) Pools and productivity calculated at equilibrium
(2) Pools and productivity calculated for initial years after beginning of fertilizer application
(3) Utilized vegetation as pasture.

are representative of the range of yields obtained by Bedouin farmers in the region even in good years. With the addition of 90 kg N per ha as fertilizer, the N-limited grain yields for equivalent systems are 3.5 to 4.0 ton per ha, again equivalent to the yields obtained in the region in good years on intensively managed croplands. The equilibrium levels of organic N in the soil of the unfertilized systems (1.35 to 3.00 ton per ha) are also within the measured range. With the addition of 90 kg N per ha as fertilizer, the organic N is expected to increase at rates of 19 to 34 kg per ha per year. It is predicted that in an 'unmanaged ecosystem', biomass production would arrive at an equilibrium level of 5.0 ton per ha and soil organic N at 8.0 ton per ha. If grazed more heavily by domestic ruminants, equilibrium biomass production would be 3.9 ton per ha and animal production would be 53 kg liveweight per ha, also a realistic figure. Thus, even though the model used is very simple, it appears to be robust and gives a recognizable picture of the known aspects of the N balance in the region.

3.3.4. *Experimental study of some nitrogen transformations in a semi-arid environment*

One of the main specific assumptions of the equilibrium model was that there is negligible N loss from the soil by leaching, volatilization or denitrification.

Table 3–2.
Cumulative percent recovery of ^{15}N labelled fertilizer N in soil and plant during three consecutive seasons at the Migda Experimental Farm, Israel (Source: Feigenbaum et al., 1984).

Fertilizer treatment (kg N ha^{-1})	Plot (1)	Number of growing seasons	Plant Shoot	Root	Soil Mineral	Organic	Total recovery (%)
A(2)	micro						
60		1	29.2	1.6	33.8	10.9	75.5
60+60		2	45.1	0.7	2.2	8.0	56.0
60+60+60		3	36.4	0.4	0.9	6.9	44.6
B	micro						
180		1	32.1	1.8	55.8	7.9	97.6
180		2	61.0	<1	1.0	7.7	69.7
180		3	64.9	+	1.0	7.2	73.1
B	open						
180		1	21.9	1.1	48.6	12.5	84.1
180		2	29.3	0.3	0.1	8.0	37.7
180		3	36.6	0.1	0.2	8.4	45.

(1) Microplots enclosed by hard plastic tubes, 0.3 m internal diameter, sunk to a depth of 0.6 m into the soil; open plots were 1.25 x 1.25 m
(2) Fertilizer treatment A: 180 kg N ha^{-1} applied as KNO_3 given in three split applications of 60 kg N ha^{-1} each season; fertilizer treatment B: 180 kg N ha^{-1} given as a single application at the beginning of the experiment.

Loss from faeces and urine in the grazed systems was assumed to be fairly high. These assumptions were tested in a series of experiments with labelled nitrogen, ^{15}N, at the Migda Experimental Farm (Feigenbaum *et al.*, 1984). Most of the labelled N could be accounted for by the end of the first year of the experiment. This was a relatively dry year (200 mm rainfall) and both leaching and waterlogging were minimal. Nevertheless, between 2.4 and 24.5 percent of the applied N could not be accounted for in the small enclosed microplots used in the experiment and 15.95 percent in open plots (Table 3–2). Some of the discrepancy could have been due to measurement error, but it is also possible that some of the applied N was leached below the 60 cm that were sampled. A large amount of the residual mineral ^{15}N was taken up by the crop in the following growing season, but by now 30 to 44 percent in the microplots and 62 percent in the open plots could not be accounted for. The second year was a much wetter year (342 mm) and there was leaching beyond 180 cm depth. At the end of the second growing season there were only very small amounts of labelled N in the soil mineral fraction, but relatively much more in the soil

Table 3–3.
Conservation of mineral nitrogen in a fallow soil in a semi-arid region (Source: Seligman et al., 1985).

^{15}N Component			Percent recovery of applied	
			Wheat	Fallow
Plant	–	1979	23.0 ± 1.7	–
		1980	6.0 ± 2.0	–
		total	29.6 ± 2.4	–
Soil	–	mineral N, 1979	48.6 ± 43.2	61.1 ± 3.2
		1980	0.1	1.5
		organic N, 1979	12.5 ± 7.2	2.0 ± 0.7
		1980	8.0 ± 5.3	5.7 ± 1.3
Total recovery at end of season				
		1979	84.1 ± 5.3	63.1 ± 2.6
		1980	37.7 ± 2.3	7.2 ± 1.2

organic N fraction. By the end of the third year, the amount of labelled N in the soil N pools hardly changed but there was additional uptake of labelled N by the crop. At that stage, 44 to 73 percent of the applied N could be accounted for in the microplots and only about 45 percent in the open plots (Table 3–2). Most of the loss occurred during the second, wet year at the end of which the soil moisture content in the deepest measured layer (150–180 cm) approached field capacity (Seligman *et al.*, 1985). The loss of ^{15}N from the 0–180 cm soil layer in the wet year was even more dramatic in an area maintained as a fallow. Here there was no uptake by plants at the end of the first season and only 63 percent of the applied N could be accounted for in the 0–60 cm layer (Table 3–3). Some may have leached into the deeper soil layers, but by the end of the second year, down to a depth of 180 cm, only 2 percent of the applied N could be detected in the soil mineral fraction and about 6 percent in the soil organic N fraction. Most of the residual mineral N from the previous year appeared to have been lost by leaching. At the deepest measured layer (150–180 cm) the soil moisture content in both the fallow and cropped treatments was high and, from the shape of the soil moisture profile, it appeared that at that depth there was very little soil moisture extraction by the roots. In a previous wet year, the root systems of both wheat and native vegetation appeared to be active down to 180 cm at least (van Keulen, 1975). The difference could be related to the development and maturation of the wheat in relation to the distribution of water availability over the season.

It must be concluded that in the wetter years there can be serious loss of mineral N from the soil by leaching and possibly by denitrification too. The question remains how these losses tally with the 'realistic' predictions of the N balance model that assumed that there were no such losses from the soil. The contradiction may be more apparent than real because wet years with moisture penetration below 180 cm are more the exception than the rule in semi-arid environments. In addition, the amount of N in the soil mineral pool is generally much smaller than under the heavy N applications used in the experiment. Both these factors would reduce the importance of these losses in the long-term equilibrium situation.

Other assumptions of the simple model concern the values given to the parameters that control the flow rates of N from the plant residues to the mineral and to the organic soil N pools. The fraction in the residues that was mineralized within 1–1.5 years was set at 0.6 and the remainder was transferred to the organic N pool. This parameter was subsequently estimated experimentally by using residues of plants grown on a nutrient solution containing [15]N (Seligman *et al.*, 1986). These were incorporated into the top 15 cm of soil in enclosed microplots. Over the next three years, wheat was grown in these plots and the uptake of ^{15}N by the crop was monitored. In the course of the three years, 6 to 16 percent of the ^{15}N applied in the organic residues was recovered in the vegetation. This is similar to the fraction recovered from ^{15}N labelled leguminous residues in Australia (Ladd, 1981). It appears then that 60 percent transfer of residue N to the mineral pool is a gross overestimation. However, this would mean that the flow from the residues to the organic N fraction was underestimated. So, it is possible that in the equilibrium situation similar amounts of mineral can be taken up, because the smaller flow from the residues would be partially compensated by greater flow from the 'stable' organic N pool to the mineral N pool.

These theoretical and experimental studies help to give a quantitative picture of the overall aspects of the N balance in a semi-arid environment, but do not yet allow one to gain insight into the interactions between nitrogen and water relations in a given crop or vegetation stand. In order to study the intra-annual dynamics of nitrogen in a soil-plant system, a more detailed model is necessary that can cover the main N transformations, water relations and plant growth processes in a crop or pasture. This will be discussed in the following subchapters.

3.4. **Plant production limited by weather, mainly water availability**

3.4.1. *The relation between biomass production and water use*

With decreasing water supply to the shoot, leaf tissue dehydrates, stomata close and CO_2 cannot be taken up so that assimilation and growth cease. The interdependence between water use and dry matter production was recognized early in agricultural research and many experiments have been carried out to determine the exact quantitative relations between the two. Much of the earlier

38

work, carried out in the US was reviewed by Briggs and Shantz (1913) and Kiesselbach (1916). A classic study was the analysis by de Wit (1958). Since then excellent reviews have been published by Arkley (1963), Hanks (1974), Tanner and Sinclair (1983) and others. In discussing water use by plants a distinction is drawn between transpiration efficiency, *i.e.* the amount of carbon dioxide fixed or the amount of dry matter produced per unit of water actually transpired by the vegetation, and the water use efficiency, *i.e.* the amount of dry matter produced per unit input of water into the soil-plant system. The first is a measure of plant physiological efficiency, while the latter, which includes losses of water due to deep drainage and soil surface evaporation is a measure of crop performance in relation to water supply.

3.4.1.1. *Transpiration efficiency*

Exchange of water vapour and CO_2 between the leaf and the atmosphere takes place along the same pathway and is governed by the same processes of diffusion and turbulence. As a consequence there is a close relationship between transpiration and assimilation, particularly within species (de Wit, 1958), but

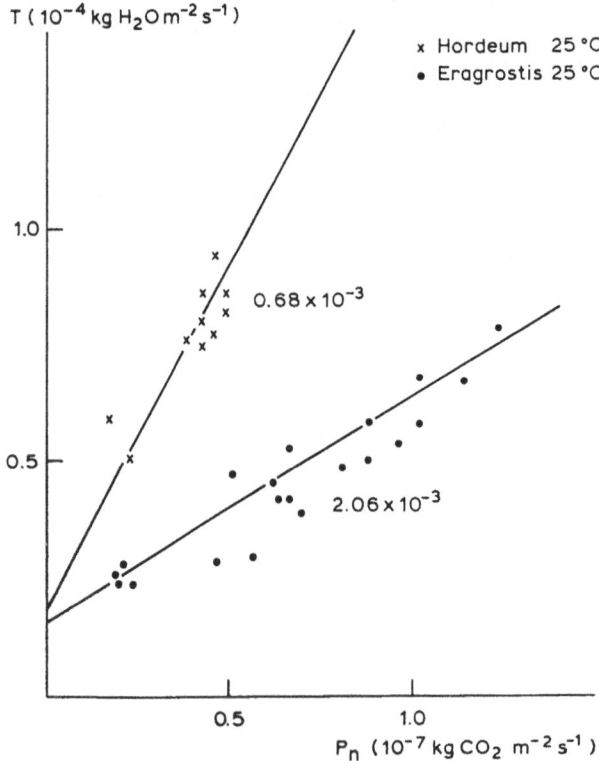

Fig. 3-2. The relation between the rate of net photosynthesis (Pn) and the rate of transpiration (T) for attached individual leaves of Hordeum murinum (C$_3$) and Eragrostis tremula (C$_4$), measured under controlled conditions.

the relationship is not one because the driving forces for the two processes are different. The gradient in water vapour concentration drives transpiration and the gradient in CO_2 concentration drives assimilation. The proportionality between these processes varies under different environmental conditions and with different plant species.

Differences between species are illustrated in Figure 3-2. *Hordeum murinum* L. is a C_3 species from the natural vegetation in a mediterranean environment (Lof, 1976) and *Eragrostis tremula* Chst. is a C_4 species from the natural vegetation in a sahelian environment (Cissé and Breman, 1982). The data refer to individual leaves of intact plants, grown and measured under controlled
conditions in Wageningen. Ambient temperature (25 °C) and air humidity for both species were similar and the variation in net assimilation and transpiration was achieved by varying the radiation intensity during the measurements. Average transpiration efficiency of the C_4 species was 20.6 g CO_2 fixed per kg of water transpired, against a value of 6.8 for the C_3 species. This difference can be traced to the difference in substomatal CO_2 concentration between the two plant types, which was about 250 vppm in the C_3 species in this experiment and about 120 vppm in the C_4 species. As a result, the gradient in CO_2

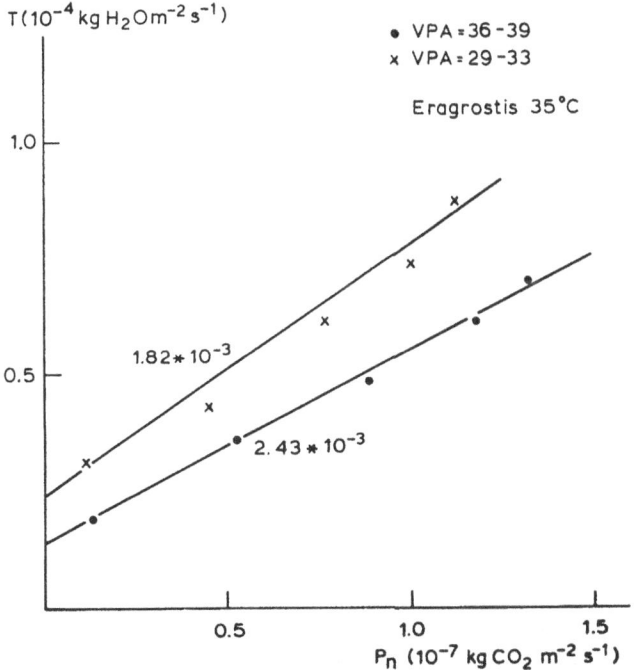

Fig. 3-3. The relation between the rate of net photosynthesis (Pn) and the rate of transpiration (T) for attached individual leaves of Eragrostis tremula (C_4), measured under controlled conditions at two ranges of air vapour pressure (VPA in mbar).

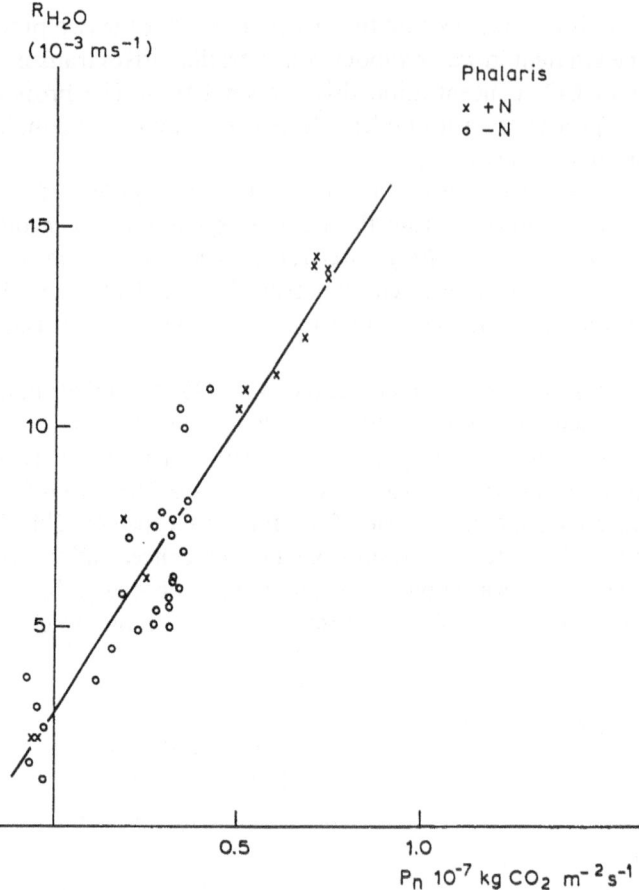

Fig. 3-4. The relation between the rate of net photosynthesis (P_n) and total resistance for water vapour transfer (R_{H2O}) for attached individual leaves of Phalaris minor, measured under controlled conditions at two levels of N supply.

concentration between the outside atmosphere (about 350 vppm in these experiments) and the substomatal cavity was much greater for the C_4 species and the rate of influx of CO_2 was consequently much higher in the C_4 species even though stomatal opening, and thus transpiration rates were the same.

The effect of environmental conditions is illustrated in Figure 3–3; *Eragrostis tremula* Chst. was maintained at 35 °C at an actual vapour pressure in the ambient air of about 30 mbar in one experiment, and at about 38 mbar in a second experiment. The difference in water vapour gradient produced a transpiration efficiency of 18.2 g CO_2 fixed per kg water transpired in the low humidity situation compared to 24.3 g in the high humidity situation. These results illustrate the point that in regions with summer rainfall, where humidity is generally low during the growing period, the transpiration efficiency for plants with comparable photosynthetic systems, will be lower than in regions with winter rainfall, with much higher humidities during the growing period.

The interesting effect of nutritional status on water use efficiency is shown in Figure 3-4. The data are from an experiment with *Phalaris minor* L. in which half of the plants where transferred to a nitrogen free solution after some weeks of growth, while the other half continued to receive full nutrient solution (Lof, 1976). The plants from which nitrogen was withheld showed clear symptoms of nitrogen deficiency and Figure 3-4 shows that their photosynthetic capacity was impaired. Even so, all the data points fell on the same straight line that relates net photosynthesis to the total conductance for water vapour, *i.e.* the inverse of the sum of stomatal and boundary layer resistance. This indicates that there is no difference in transpiration efficiency between the plants well-supplied with nitrogen and the nitrogen-deficient plants. Similar observations were made on maize plants by Goudriaan and van Keulen (1979) and Wong *et al.* (1979), but Bolton and Brown (1980) found a substantial decrease in transpiration efficiency with decreasing nitrogen content in tall fescue and *Panicum milioides* Nees. The reason for this difference in behaviour is not clear, but it may be related to differences in stomatal control.

3.4.1.2. *Water use efficiency*
In the field, evaporation from the soil surface and drainage of water beyond the potential rooting zone can vary from year to year, from place to place, and in some situations also between treatments, so that even if transpiration efficiency was constant for a given set of conditions, the water use efficiency can vary. An example (Table 3-4) is an experiment with alfalfa conducted in North Dakota (Bauder and Bauer, 1978). There, water use efficiency increased with increasing rate of irrigation, because non-productive water loss from soil surface evaporation, the intercept with the x-axis, was a smaller proportion of total water use at higher irrigation rates (Fig. 3-5).

Another example is an irrigation experiment with potatoes on the rainless pampa de La Joya in Southern Peru (Versteeg, 1985). In Table 3-5, the water use efficiency is presented for two cultivars and five irrigation treatments. For cv. Revolución there was an optimum at an intermediate level of water application, while for cv. Desirée, with water use efficiencies more than a factor

Table 3-4.
Forage yield, water use and water use efficiency for an irrigated alfalfa crop.

Treatment	Forage yield (kg per ha)	Water use (mm)	Water use efficiency (g per kg)
W1	4,324	363	1.19
W2	7,536	561	1.34
W3	8,994	635	1.42
W4	8,747	643	1.36

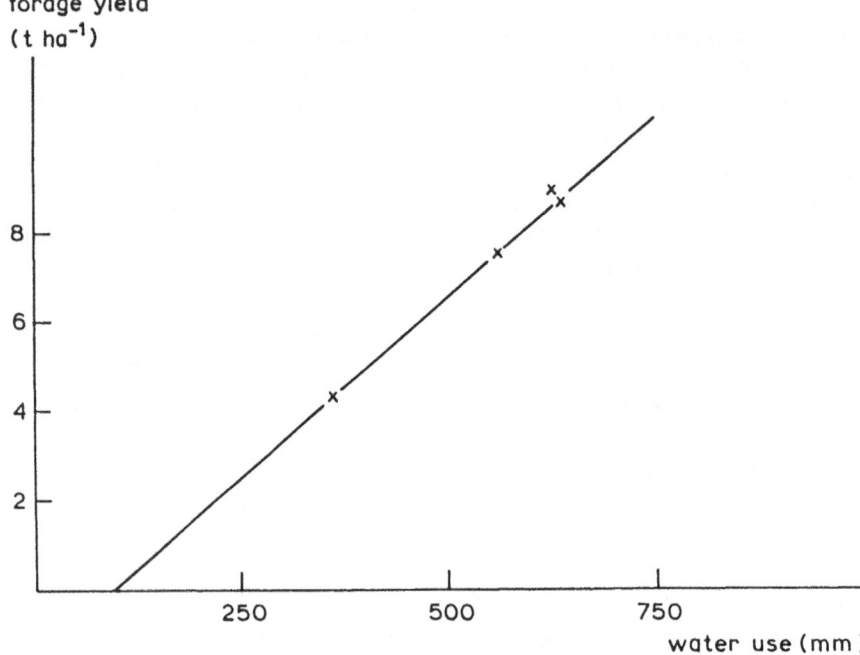

forage yield
(t ha⁻¹)

water use (mm)

Fig. 3–5. The relation between total water use and forage yield for alfalfa grown under irrigation in North Dakota (Source: Bauder and Bauer, 1978).

two lower at the lower application rates, the values increased monotonically with increasing application rate, even though the transpiration efficiencies of both cultivars were practically the same at 4.5 and 4.2 g dry matter per kg water for Desirée and Revolución, respectively (Fig. 3–6). The non-productive water loss (the intercept with the x-axis) on the other hand is about 150 mm higher for cv. Desirée because of its slower initial growth.

Table 3–5.
Dry matter yield (haulm and tubers), total irrigation and water use efficiency for a potato crop.

| Total irrigation | cv. Revolución | | cv. Desirée | |
| | Dry matter production | Water use efficiency | Dry matter production | Water use efficiency |
mm	(kg per ha)	(g per kg)	(kg per ha)	(g per kg)
690	8,600	1.25	3,600	0.52
760	12,600	1.66	6,600	0.68
840	15,900	1.89	9,800	1.17
970	17,300	1.78	12,900	1.33
1,200	17,700	1.47	16,250	1.35

Water use efficiency in terms of total input of moisture in the system is important to the farmer, especially under irrigated conditions, but in itself cannot explain important differences between species or even cultivars of the same species. This is also the case in many arid and semi-arid regions, where the water use efficiency of the natural vegetation expressed in terms of dry biomass production per unit precipitation is extremely variable and often very low, because of large losses from both soil surface evaporation and runoff. Le Houérou and Hoste (1977) used 45 data points from the Mediterranean Basin to calculate the relationship between annual precipitation and annual rangeland production. The slope of the calculated regression line was 8.68 kg per ha per mm, indicating a water use efficiency of $8.68*10^{-4}$ kg dry matter per kg water. The intercept with the x-axis at 48 mm annual precipitation, gives an estimate of the non-productive water loss, which in this situation would be practically all soil surface evaporation. Similar water use efficiencies were found in the arid soudano-sahelian region (100–400 mm) by Breman (1975), 0.9 g per kg, and in the South African 'veld' by Walter and Volk (1954), 0.8 g per kg. Where rainfall is higher and runoff is important, the relationship between water use and rainfall is not linear. In such cases the linear relationship gives a misleading (and irrelevant) water use efficiency of 0.24 g per kg (Breman, 1975).

All these values are substantially lower than those reported in Tables 3–4 and 3–5. The main reason is probably that production in natural rangelands is not determined so much by moisture availability during the rainy season as by the availability of nutrient elements, especially nitrogen and phosphorus (Penning de Vries and van Keulen, 1982; Power, 1980a, 1980b; van Keulen, 1975). The

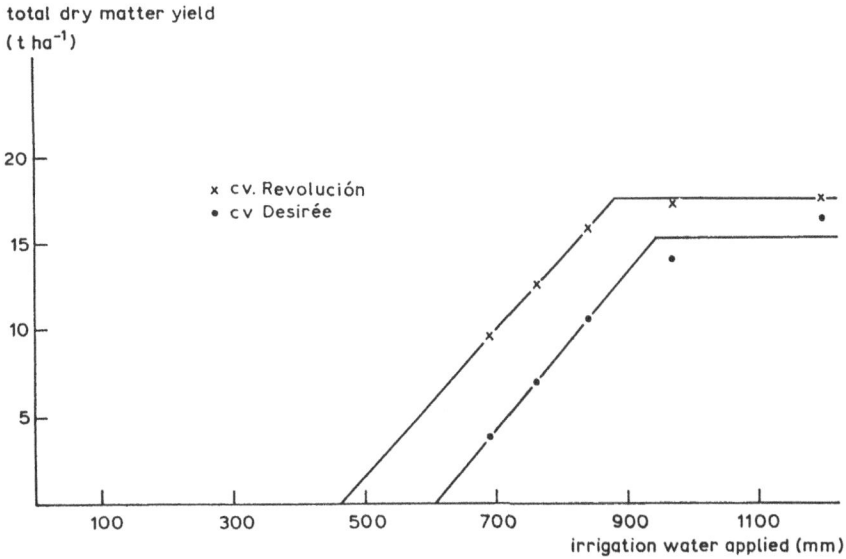

Fig. 3-6. The relation between total water use and total dry matter yield for two potato varieties grown under irrigation in Peru (Source: Versteeg, 1985).

consequent reduction in plant growth and greater soil exposure, could greatly increase soil surface evaporation and runoff losses, and so drastically reduce water use efficiency.

3.4.2. *The simulation model ARID CROP*

ARID CROP (van Keulen *et al.*, 1981; van Keulen, 1975) is a simulation model based on the soil moisture balance and on plant growth processes. It calculates the time course of dry matter production in a stand of annual plant species and the distribution of moisture in the soil below the plant canopy. It simulates potential water-limited growth and ignores other limiting factors like nutrient deficiencies, weeds, pests and diseases. Inputs required for the model are daily observations from standard meteorological stations, and specifications of vegetation characteristics that are related to photosynthetic performance, distribution of assimilates among various plant organs, leaf weight/area ratio, and root development. Soil physical data are needed that specify the soil moisture content at field capacity, wilting point and under air-dry conditions, respectively. The model is based on time intervals of one day and this resolution determines the formulation of all relevant processes.

3.4.2.1. *Short description of the model*
Soil physical processes
For the description of the physical processes in the soil, the total profile is divided into an arbitrary number of compartments (or layers), varying in thickness, each of which is considered homogeneous. The top compartments are divided into thinner layers than the bottom compartments so as to increase the sensitivity of the water balance to the important processes that take place near the soil surface. The properties of the various compartments are not necessarily the same, so that layered profiles can be accomodated.

The total amount of water infiltrating into the soil, is determined by precipitation, irrigation, run-off or run-on. Once in the soil, the water cascades through the various soil compartments filling each one to field capacity from the top downwards until all the water is dissipated. Any surplus after the lowest compartment attains field capacity is lost to deep drainage.

Potential evaporation from the soil surface is calculated by a Penman-type equation which partitions the incoming energy between the vegetation and the bare soil. The current rate of water loss from the soil surface is also dependent on the dryness of the top soil compartments. Transport between soil compartments along developing potential gradients that result from uneven drying is not described in the model. Instead, a descriptive method is used to 'mimic' the redistribution process with parameters derived from a detailed physically based model (Stroosnijder, 1982; van Keulen, 1975). The total calculated water loss through soil surface evaporation is withdrawn from the various compartments as a function of soil physical properties and the current moisture distribution in the soil profile.

Soil temperature is calculated as a running ten-day average of air temperature. This is a rather crude approximation of a complex heat exchange process but it does represent both the lag effects and the damping effects that differentiate soil temperature from air temperature. The precision of this approximation is sufficient for the time resolution of the model.

Uptake of moisture from the soil by plant roots is driven by potential crop transpiration, but is modified by root and moisture distribution in the profile as well as by root conductivity and water viscosity, both of which are dependent on soil temperature. Allowance is made for partial compensation whereby water uptake by the root is preferentially greater in moister soil layers.

Growth of the vegetation
The effect of species composition on the water relations and growth of a mixed stand of annual species is generally small as long as they belong to either the C_3 or C_4 photosynthesis groups, which is usually the case. Consequently, the vegetation is considered as a botanically homogeneous stand. Germination is treated simply and starts when the soil moisture content in the upper 10 cm of the soil is above wilting point. Establishment of the seedlings occurs when the accumulated soil temperature sum exceeds 150 day-degrees (above zero). If the soil dries out before germination is completed, the seedlings die. A new wave of germination can be initiated after rewetting. The size of the germinating seed stock in the soil is defined as an input variable and determines the value of the biomass at establishment. It is difficult to quantify the seed stock and the relation between seed weight and initial leaf biomass, and so it is necessary to calculate the initial biomass value from field data or to use a carefully estimated constant value.

Vegetation growth is calculated as a product of the actual rate of transpiration times the transpiration efficiency, on the assumption that growth, as a rule, is closely related to transpiration. Transpiration efficiency is defined in this case as the ratio between potential growth rate and potential rate of transpiration and is assumed to be independent of both soil moisture conditions and development stage of the vegetation. The first assumption may not be true, as there are differences between species in their response to restricted water supply. In general, that is not important quantitatively, because the actual amounts of water transpired during periods of severe moisture stress are small. These small differences can, however, determine the survival and regrowth of plants after a long dry spell and in such cases can be critical. Our knowledge of survival physiology is, however, too fragmentary to allow an accurate quantitative description of these processes. The second assumption is probably valid in most practical conditions as long as nutrients are not severely limiting and the proportion of old senescing leaves is small.

The potential rate of transpiration is calculated from radiation intensity and the combined effect of wind speed and humidity, using the combination method of Penman. The empirical relations that are used to derive potential transpiration from daily meteorological data are based on a detailed model of canopy transpiration (de Wit *et al.*, 1978).

The potential rate of growth of the vegetation is derived from the gross rate of canopy assimilation, taking into account the losses incurred by maintenance respiration and growth respiration. Gross assimilation is calculated with an algorithm developed by Goudriaan (1986), as a function of incoming radiation intensity, green leaf area index of the vegetation, optical properties of the leaves and their photosynthetic characteristics. The latter are represented by the photosynthesis-light response curve, which is characterized by the light use efficiency at the light compensation point and the maximum rate of assimilation at light saturation (de Wit, 1965). Both characteristics are influenced by the moisture status of the vegetation, and are adversely affected by prolonged moisture stress. The maximum rate of assimilation is also influenced by temperature, albeit with a rather flat optimum (de Wit *et al.*, 1978) and is reduced towards the end of development to account for the effect of translocation of nutrients, particularly nitrogen, from the vegetative tissue to the growing seeds. Maintenance respiration, which is the energy expense for biological functioning of the existing plant components is calculated from the weight of the living biomass and the prevailing temperature. Growth respiration, associated with transport of assimilates and the conversion of primary assimilation products into structural plant material, is described in the model by a constant conversion efficiency.

Partitioning of the total increase in dry weight among the leaf blades, stems, roots and seeds is dependent on the phenological development stage and the moisture status of the vegetation. As moisture deficiency increases, a larger proportion of the assimilates is allocated to the root system.

Plant tissue can die from either water shortage or from senescence. It is assumed that even with fully closed stomata some water loss from the plants occurs through the cuticle. When the actual rate of water uptake cannot compensate for that water loss, part of the plant tissue dehydrates and dies at a rate proportional to the difference between water loss and water uptake, with an assumed time constant of five days to account for the buffering capacity of the plants. Leaves have a limited life span so that after some time, some of the older leaves senesce and die. Death rate of leaf tissue increases after flowering as vital materials, mainly minerals, are translocated from the leaves to the developing seeds. In the model, it is assumed that death from senescence is negligible until flowering and only starts after the onset of seed-fill.

The phenological development pattern of the vegetation is characterized by the rate and order of appearance of vegetative and reproductive plant organs. The order of appearance is a species characteristic and in most cases is independent of environmental conditions. The rate of development, however, is strongly influenced by those conditions, notably temperature and daylength. In the model the rate of development is defined as a linear function of temperature with a threshold value of 3.75 °C. Its dimension is d^{-1} so that the integrated value of the rate of development is dimensionless and defines the development stage. Its value is 0 at emergence, 0.65 when seed fill starts and 1 at physiological maturity. Annual plants germinating after December 21 in a mediterranean environment accumulate less day-degrees till flowering and maturity than plants germinating before that date (Ungar, unpublished data). The effect of daylength is not treated explicitly but exits through the relationship between daylength and temperature. As a consequence, with later germination, maturity is attained more rapidly.

The morphological development of the vegetation is limited to a description of leaf area development that is derived from the increase in leaf weight, by using a temperature-dependent specific leaf area factor.

The root system of the vegetation is defined by its mass and its vertical extension. It is assumed that a 'root front' moves downward in the soil at a rate determined by soil temperature. Horizontal gradients in root density are not taken into account, because it is assumed that root density is always sufficient to supply the required water uptake (van Keulen et al., 1975). The water uptake is dependent on the average moisture content in each soil compartment and not on potential gradients developing around individual roots.

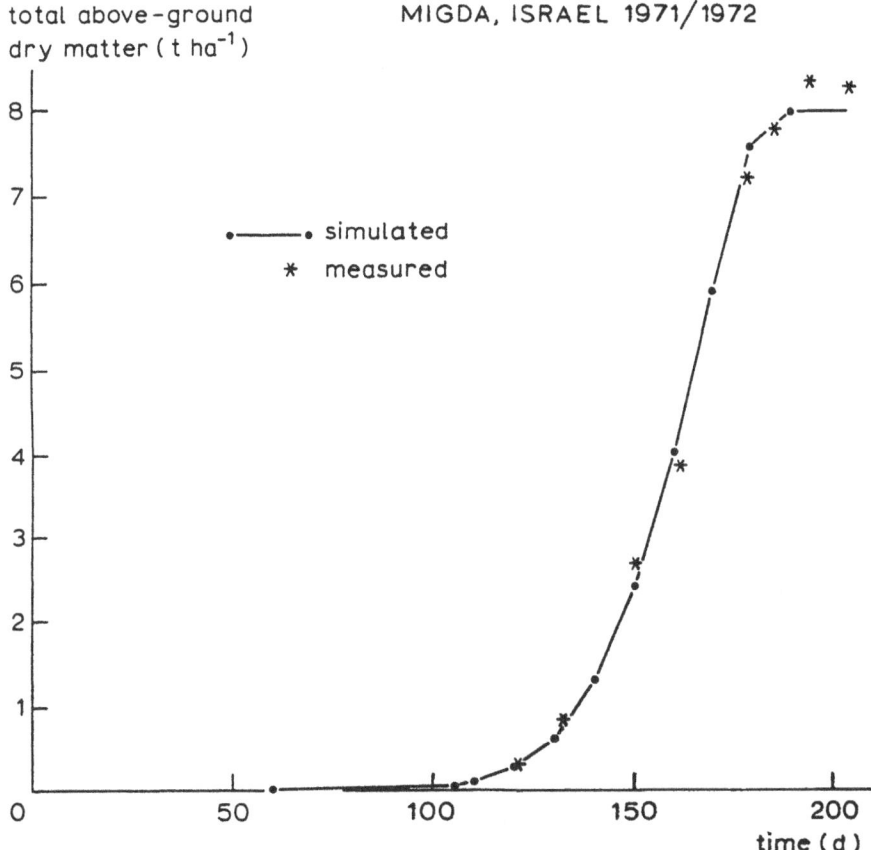

(a)

48

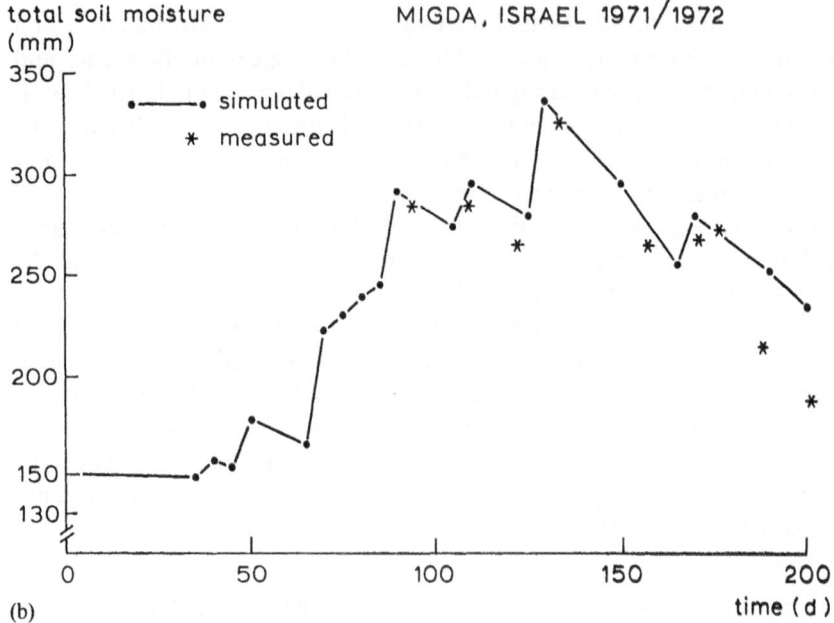

Fig. 3-7. Comparison of measured and simulated total dry matter accumulation (a) and total soil moisture (b) for natural vegetation in Migda, Israel in 1971/1972.

3.4.2.2. *Validation of the model*

In the period 1970–1980 extensive experimentation was carried out at the Migda Experimental Farm as part of a joint Dutch-Israeli research project titled 'Actual and potential production of semi-arid grasslands' (van Keulen *et al.*, 1981). Growth of fertilized natural pasture was monitored throughout the growing period, as was the water balance in the soil. Some of the data collected have been used to develop and calibrate the model and the rest for model validation. In this section the results are discussed for three growing seasons, that are fairly representative of conditions in the better rainfall years in the northern Negev and give a good indication of model performance.

Detailed observations were made first in the 1971/1972 growing season and consequently these results have been used over the years to calibrate model performance. This was a good season for the region with annual precipitation well above average (350 mm *versus* 250 mm) and a very favourable distribution. The simulated growth curve follows the observed values very closely (Fig. 3–7a) and is always well within one standard deviation from the observations. Special attention is drawn to the fact that early development of the vegetation was very slow, with only 350 kg per ha present 60 days after emergence. The simulated growth curve is sensitive to initial biomass, which was calibrated to a low value of 0.75 kg per ha. Higher values cause the simulated growth curve to shift strongly to the left. This low initial biomass occurred because the field was disked prior to germination to incorporate the ammoniacal fertilizer. Many

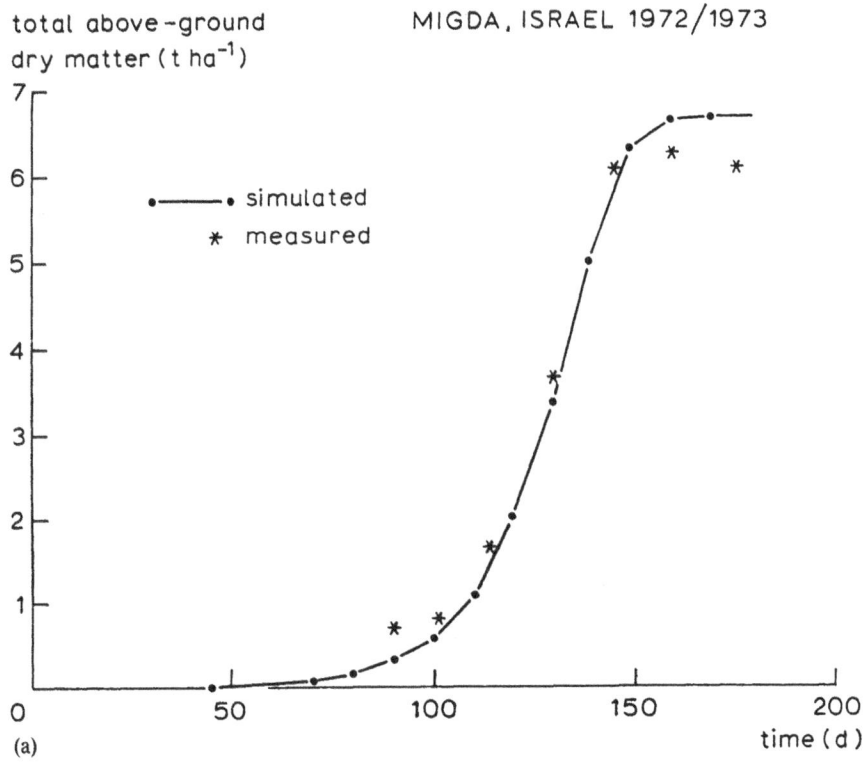

(a)

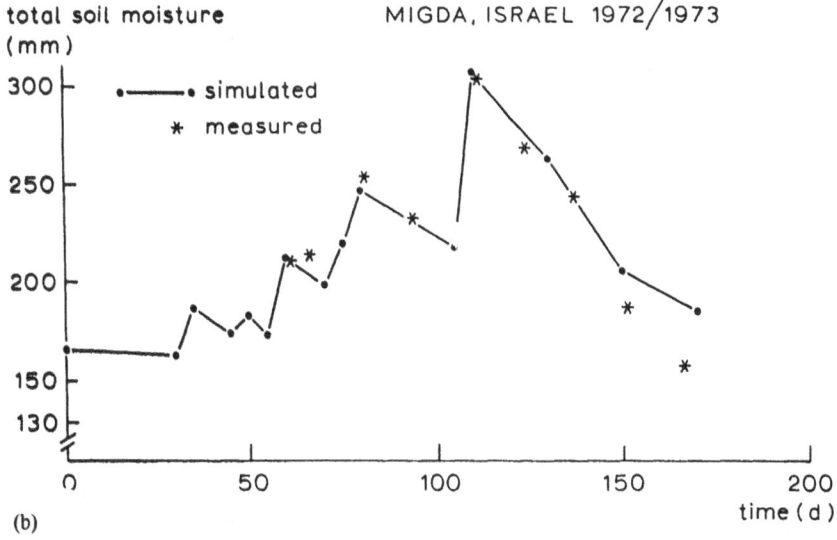

(b)

Fig. 3-8. Comparison of measured and simulated total dry matter accumulation (a) and total soil moisture (b) for natural vegetation in Migda, Israel in 1972/1973.

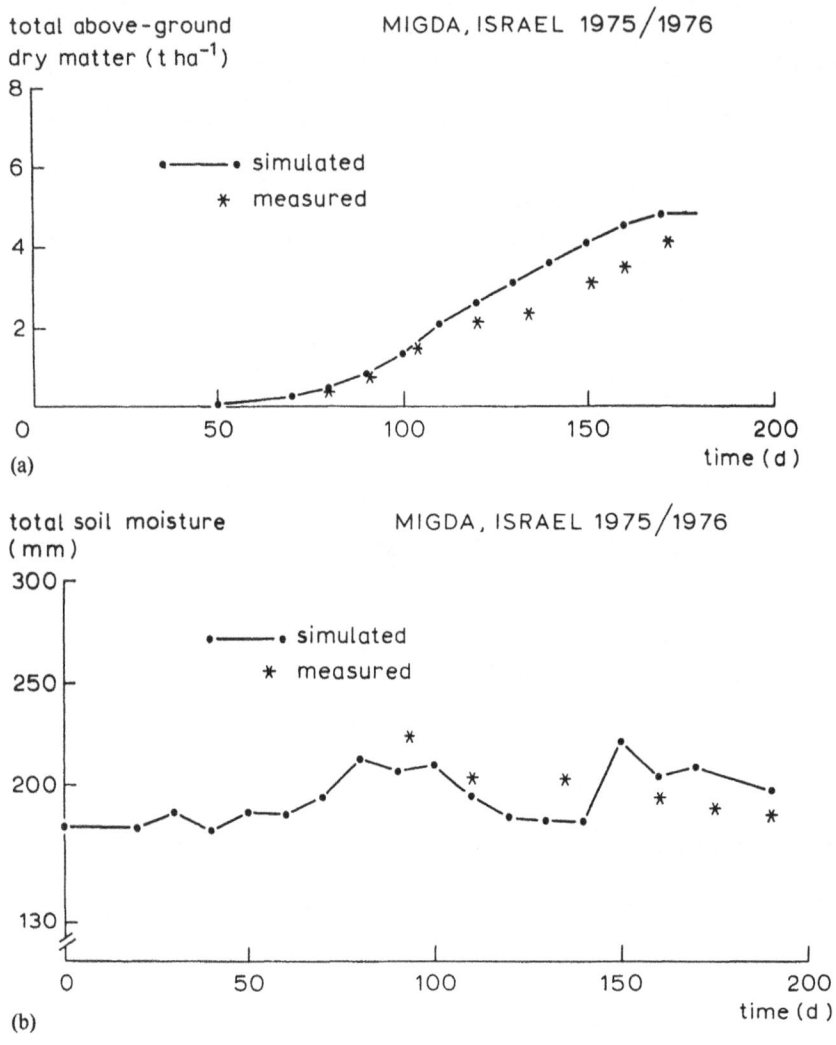

Fig. 3-9. Comparison of measured and simulated total dry matter accumulation (a) and total soil moisture (b) for natural vegetation in Migda, Israel in 1975/1976.

seeds were buried in the operation, so that the density of germinating seedlings was very low.

Soil moisture dynamics over the growing season were simulated fairly accurately for most of the season (Fig. 3-7b), with deviations increasing towards the end of the growing season. It is possible that in the field most of the late light rains did not reach the soil but were intercepted by the heavy stand of vegetation. As the 1971/72 data were used for calibrating the model, they cannot be used for 'validation'. This requires an independent set of data (van Keulen, 1976).

In the 1972/1973 growing season, rainfall was about average at 245 mm and the distribution was again favourable. The simulated growth curve for this season, based on an initial biomass of 5 kg per ha, again follows the observations closely for most of the season (Fig. 3–8a), with growth slightly overestimated towards the end. Soil moisture dynamics was also simulated accurately for most of the growing period, but with deviations towards the end of the season (Fig. 3–8b). Between day 150 and day 170, when the growth rate was overestimated, the model *underestimated* moisture extraction from the profile. This again could be caused by interception effects, as the model assumes that all of the 19 mm of rainfall between day 150 and day 160 infiltrated completely into the soil. However, no hard evidence for interception loss is available.

The 1975/1976 season is an example of a relatively dry season. Total precipitation (204 mm) was not particularly low, but the distribution was unfavourable. Early rains, which initiated germination were followed by a long dry spell before the rains resumed. The simulated growth curve, based on an initial biomass of 50 kg per ha, follows the observations closely during early growth (Fig. 3–9a). However, during the long dry spell between day 100 and day 140, the calculated growth was overestimated by more than 500 kg per ha. During the dry spell, the cumulative relative transpiration deficit reached a high value, that in the model is assumed to reduce the potential assimilation rate because of damage to the photosynthetic machinery of the leaves. The resulting lower growth rate after resumption of the rain is well-represented in the model. Total dry matter yield reached just over 4,000 kg per ha. Simulated dynamics of soil moisture are well within the accuracy limits of the measured values (Fig. 3–9b), which, even in these homogeneous soils at the experimental site, are highly variable within a field (van Keulen, 1975).

The deviations of the model data from observations do not detract from the fact that the annual cycle of growth is simulated to a recognisable degree so that the main differences between years in the progress of growth and in the total production are reproduced quite well. In addition, not all of the discrepancies between simulated and observed results can be ascribed to model performance. In the natural mixed species pasture vegetation that characterized this study site, the heterogeneity in plant characteristics is much larger than in crop fields, with the result that sampling variation in the observed data is large. Despite these difficulties, the model does account rather well for the inter- and intraseasonal variation of biomass accumulation. With some reservation, it can be used to investigate the moisture limited grassland productivity in warm semi-arid situations.

3.4.2.3. *Application of the model.*
Calculation of long-term variability in production
Data on weather and dry matter production from an annual species pasture sward for 21 consecutive years are available for the experimental site in the northern Negev. The simulated peak dry matter yields for that period are presented in Figure 3–10, along with measured yields for both natural and

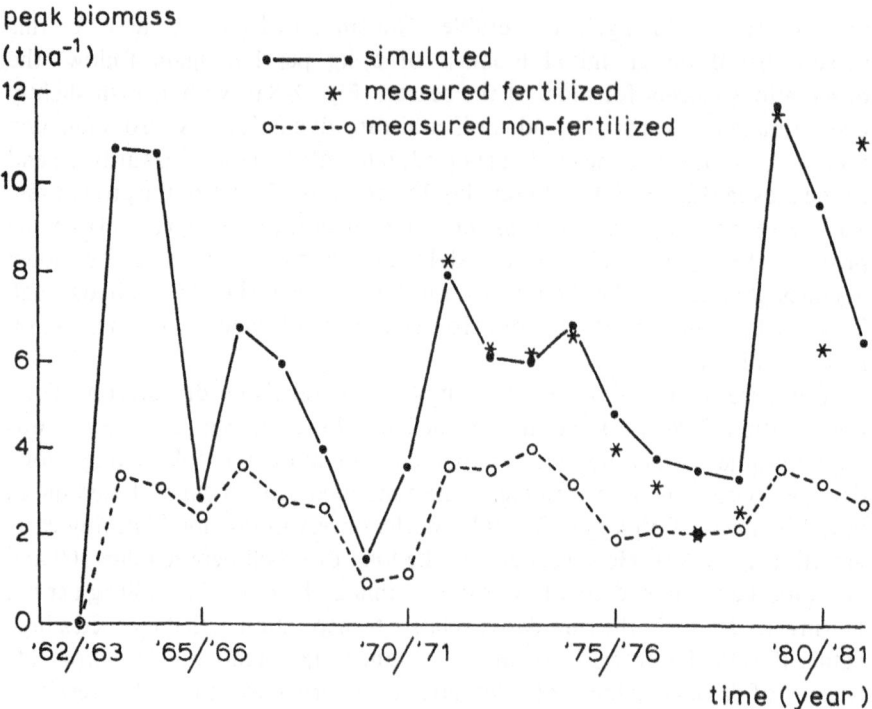

Fig. 3-10. Comparison of simulated and measured peak biomass for both fertilized and non-fertilized conditions, for natural vegetation in Migda, Israel for a twenty-one year period.

fertilized pasture swards. For the years 1962/1963 till 1970/1971 for which only peak biomass values are available, the simulations were run with an estimated initial biomass of 10 kg per ha. Whereas the course of growth is sensitive to initial biomass values, calculated peak biomass values are not (van Keulen *et al.*, 1981; van Keulen, 1975). Actual growth curves are available for the years 1971/1972 till 1982/1983 and so initial biomass values were calibrated for each season separately so that the simulation results agreed with the first measured biomass determination.

Under these conditions, when only water availability limits growth, production is extremely variable. Average calculated shoot dry matter yield for the 21-year period is 6,342 kg per ha, with a coefficient of variation (C.V.) of 65.4 percent. Average root weight over the period is 1,583 (50.7 percent) kg per ha, varying between 73 and 3,229 kg per ha. The average shoot to root ratio is 3.98 ± 1.39.

There are several reasons for the large variability in dry matter production: First, the variability in annual precipitation, which varied between 78 and 414 mm over the 21 years. Rainfall distribution contributes to the variability through its influence on the evaporation/transpiration ratio and on the timing and degree of water stress which in turn affects the distribution of dry matter in the plant. Simulated water use efficiency, *i.e.* (shoot + root)/annual rainfall

Table 3–6.
20-year water use and transpiration efficiencies of natural vegetation at Migda (simulated values)
for 1.8 ans 0.6 m. soil depth.

Variable	Range g kg^{-1}	Mean g kg^{-1}	C.V. %
Soil depth 1.8 m			
Water use efficiency (shoot+root)	0.3 – 4.05	2.72	35.5
Transpiration efficiency (shoot+root)	5.3 – 9.18	6.03	17.5
Transpiration efficiency (shoot only)	3.6 – 7.76	7.67	13.0
Soil depth 0.6 m			
water use efficiency (shoot + root)	0.3 – 3.88	2.56	38.2
Transpiration efficiency (shoot + root)	5.6 – 9.18	6.05	17.9
Transpiration efficiency (shoot only)	3.7 – 7.71	7.84	13.5

and transpiration efficiency values are given in Table 3–6. The main cause of the difference between water use efficiency and transpiration efficiency is the ratio between seasonal transpiration and seasonal precipitation. This varies between 0.06 in a very dry year with very uneven rainfall distribution and 0.62 in a relatively wet season with well-balanced rainfall distribution; the average value was 0.36 with a standard deviation of 0.14. The variation in transpiration efficiency related to total dry matter is the result of variability in environmental conditions during the growing season, notably temperature and humidity (see Section 3.3.1) which strongly influence free water evaporation during the growth period of the vegetation. The seasonal average varied between 1.8 and 3.2 mm per day over the 21-year period. A second phenomenon is that in about two thirds of the years, dry matter production calculated on the basis of moisture availability alone is substantially higher than that observed in the unfertilized natural pasture, but is close to the dry matter production in fertilized pasture. This indicates that even under these semi-arid conditions, nutrient availability is more frequently a constraint for canopy growth than is moisture availability.

Influence of soil properties
At the experimental site in the northern Negev, the soil is a deep loessial deposit with excellent physical properties for rooting. Rooting depth is only limited by the depth of water penetration. In many cases, however, natural vegetation grows on shallow soils and the presence of parent rock or other obstructions limit rooting depth. The effect of soil depth on production was examined for the same series of years as in the preceding section by assuming maximum rooting depths between 30 cm and 180 cm with 30 cm increments. The effects on average aboveground dry matter production and variability are illustrated in Figure 3–11. Average aboveground dry matter production for the 21-year

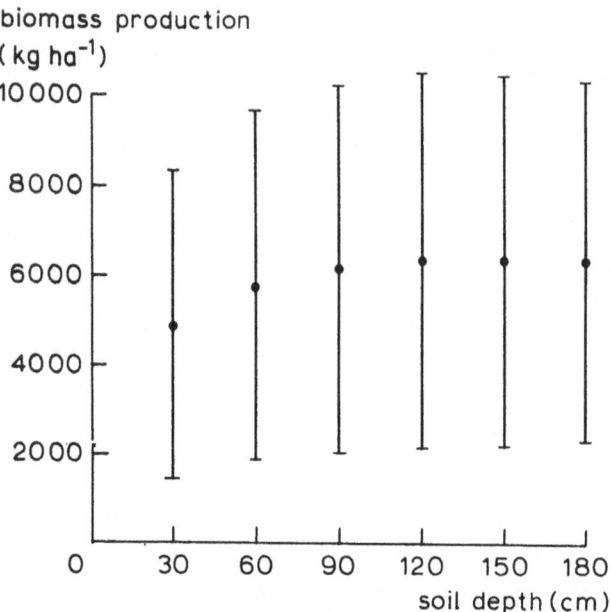

Fig. 3-11. Simulated average biomass and standard deviation for natural vegetation in Migda, Israel for a twenty-one year period at various soil depths.

period varies from 4,902 kg per ha (C.V. 70 %) for a maximum rooted depth of 30 cm to an average of 6,340 kg per ha (C.V. 65 %) at 180 cm. The main reason for the lower production at shallower soil depth is the higher proportion of non-productive water loss due to deep drainage in the case of a limited rooting depth: the average ratio of transpiration to precipitation declines. As an illustration, the water use and transpiration efficiency values for the 60 cm soil depth are given in Table 3–6. These results point to a second reason for the lower aboveground dry matter production with a limited soil depth: a somewhat higher proportion of total dry matter production is partitioned to the root system, evidently in response to more frequent occurrence of drought. This is also reflected in the lower value of the shoot/root ratio, which is here on average 3.58 (compared to 3.98 in the deeper soil).

On the whole, the effect of limited soil depth on plant production appears to be very modest. This is because the storage capacity of this loess soil is relatively high in relation to the individual rainfall events in the mediterranean environment. Where rainfall is more concentrated with larger individual showers and longer dry spells, soil depth would increase in importance.

Application in another region
A good test of the validity of the model is its performance under different environmental conditions, especially conditions that are different from those that guided the development of the model. An opportunity for such a test of the model was provided by an intensive research project in the sahelian zone of

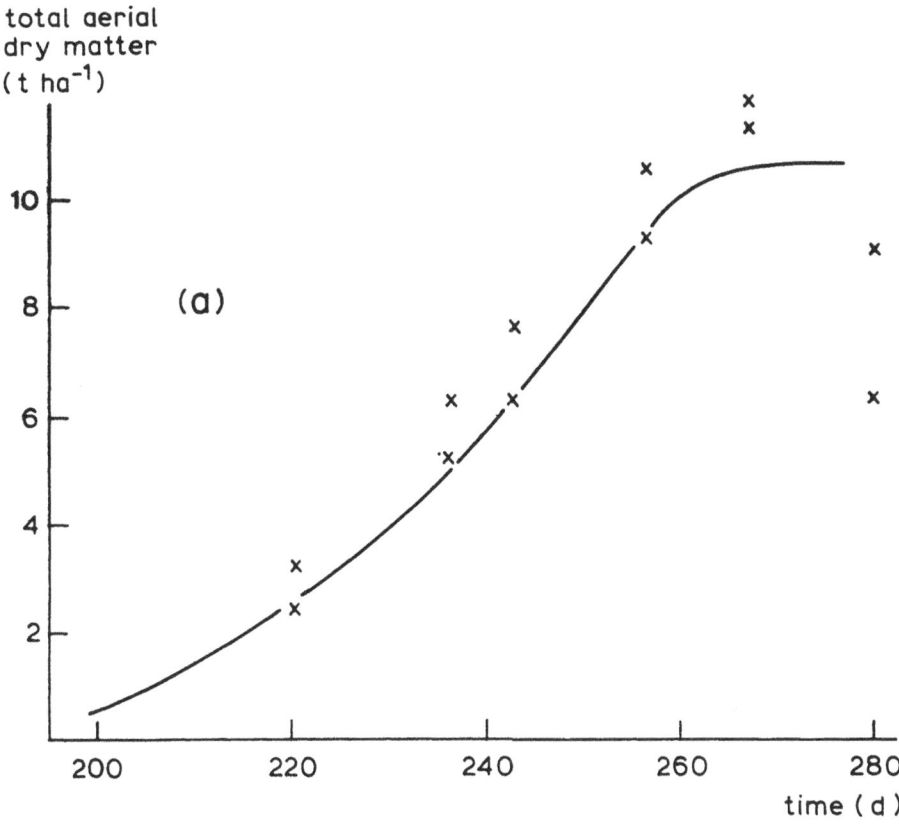

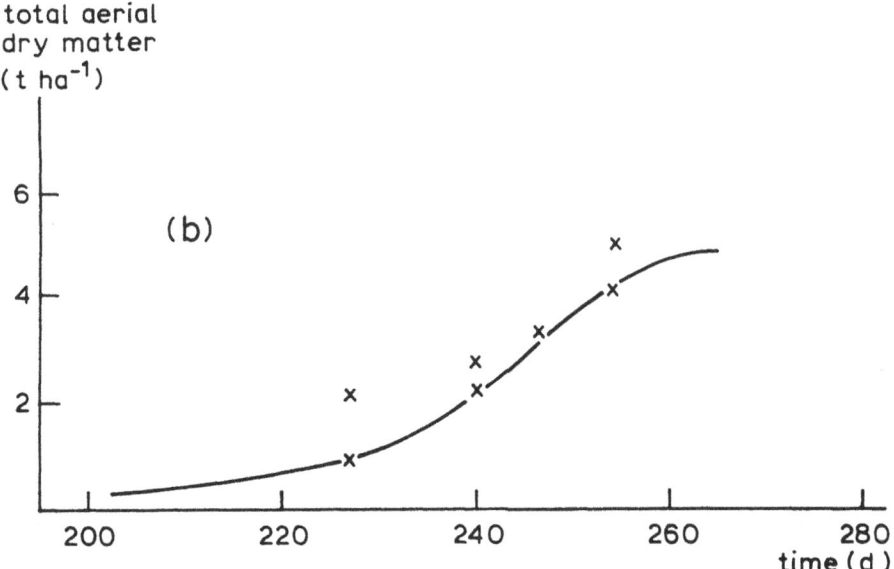

Fig. 3-12. Comparison of measured and simulated aboveground dry matter accumulation for natural vegetation in Niono, Mali, for a clay soil in 1976 (a) and a sandy soil in 1978 (b).

Mali (5° 45' E, 14° 30' N) (Penning de Vries and Djitèye, 1982). The model was used to estimate water-limited production of a seasonal annual-species sahelian grassland. The main adaptations that were necessary were related to the fact that many of the important species that compose the annual vegetation in the Sahel are C_4 species. In addition to their specific photosynthetic characteristics, their phenological development is strongly influenced by photoperiod. Another difference that had to be accounted for is related to the specific soil properties that cause heavy runoff under the high intensity rainfall common in the region (Penning de Vries, 1982). After the necessary adaptations the model performed well under these conditions. This suggests that the same basic processes that are described in the model, determine plant growth in relation to water use even under very different environmental conditions. Figure 3–12a refers to an experiment in 1976 on a clay soil with on average 13 percent run-on and a vegetation consisting of the annual grass *Diheteropogon hagerupii* Hitchc. Annual precipitation in 1976 was 587 mm, about equal to the long-term average with a fairly favourable distribution. The simulated growth curve is in close agreement with the observed data except at the very end, when senescence of the vegetation in the model is more severe than in reality. Because of the additional availability of water due to run-on, production of fertilized pasture on this site exceeded 10 tons per ha.

Figure 3–12b refers to an experiment in 1978 on a sandy soil, with on average 40 percent runoff, and a vegetation consisting of a mixture of annual grasses. Total precipitation in that season was 453 mm, well below the long-term average, and because of the heavy losses of water due to runoff, total dry matter production of this fertilized pasture was only slightly over 4,000 kg per ha. Again the simulated growth curve agrees very well with the observations over the total growing season.

The yield of natural unfertilized pasture for the two situations was 3,600 and 1,400 kg per ha, respectively. Here too, nutrient shortage during the growing season was a more serious constraint to plant production than water shortage.

3.5. Plant production limited by weather and nitrogen availability

3.5.1. *Plant growth and nitrogen supply*

Under natural conditions in the semi-arid region production is often limited by the availability of nutrients, rather than by the availability of water (Penning de Vries and Djitèye, 1982; Power, 1980a, 1980b; van Keulen, 1975; Harpaz, 1975). A comprehensive review of the main factors that should be taken into account when modelling the relationship between nitrogen and plant growth is given by van Keulen *et al.*, (1989). Here some of the aspects that have greater relevance to plant growth in semi-arid regions will be summarized.

3.5.1.1. *Uptake of nitrogen by the root system*

Plants that do not have a symbiotic relationship with N-fixing micro-organisms are dependent on mineral forms of N: NO_3-N and NH_4-N. Whereas NO_3-N moves freely with water in the soil, NH_4-N is adsorbed on the clay complex and is released to the soil solution in accordance with the laws of physical chemistry. The plant can take up ions that are brought to the root surface by mass flow with the water moving in the soil, or the ions can move along concentration gradients. The density of the root system can be important in determining the rate of flow of nitrogen to the plant especially during active growth when the concentration of ions near the root surface can be low and the replenishment of N ions must come from some distance away in the soil.

The importance of root density was studied with a simulation model based on the physical principles that govern movement of ions in water (van Keulen *et al.*, 1975). It was found that when absorbing root surfaces are 1 cm apart and there is moisture in the soil above wilting point, then, if there is a demand for N in the plant, more than 90 percent of the mineral N store in the soil will be available to the plant within 2 days by diffusion only. As most small grain crops and grassland vegetations normally have root densities greater than 1 cm per cm^3, it would appear that in most cases the root system is adequate to make all mineral soil N available to the plant. This is apparently so because the mineral N content under actively growing grass is, as a rule, very low (Theron, 1951; Theron and Haylett, 1953). Consequently, the precise measurement of root density in N-uptake studies can probably be neglected in most studies on grass or small grain agro-ecosystems.

3.5.1.2. *Nitrogen availability and photosynthesis*

The relationship between N availability and photosynthesis is not unambiguous because the many studies that have yielded many data are not always consistent. Consequently there is considerable room for interpretation (van Keulen and Seligman, 1987; Sinclair and Horie, 1989). When plotting the available data, van Keulen and Seligman (1987) found the relationship between photosynthetic rate at saturation light intensities and nitrogen content to be linear over the normal range of leaf N concentrations. Sinclair and Horie (1989) assume a curvilinear saturation relationship, with fairly far-reaching implications. There is no clear indication of a consistent relationship between N concentration in leaves and the light use efficiency at low light intensities (van Keulen and Seligman, 1987).

3.5.1.3. *Nitrogen availability and leaf area growth*

Many studies have shown that N deficiency reduces the rate of leaf growth and as a consequence, the rate of leaf area increase. This need not necessarily be directly proportional to leaf growth, because leaf thickness can vary with N supply, rate of assimilate flow and possible direct effects of light. However, the dynamics are complex and for many practical purposes, leaf area can be estimated as a linear function of leaf weight.

3.5.1.4. *Nitrogen demand of the growing vegetation*
Uptake of N from the soil depends both on the demand of N by the plant and on the availability of N in the growth medium. The demand is related to the assimilation rate of the plant and the chemical composition of the developing organs of the plant. Young tissues have high protein contents but as they differentiate, the relationship between structural carbohydrates and protoplasm changes, depending on the organ and its age. Young plants normally have high N concentrations which decrease consistently with development as a consequence of both dilution and increase of structural carbohydrate necessary to maintain the stature of a growing canopy (van Burg, 1962). The composition of the protoplasm can also change and some protein is apparently stored and not actively involved in metabolic processes (Friedrich and Huffaker, 1980) so that at equivalent development stages, the N content of plants deficient in N can be lower than that of plants adequately supplied with N. Consequently, restriction of N supply need not severely restrict growth, particularly in more mature plants (Seligman *et al.*, 1976).

3.5.1.5. *Translocation of nitrogen in the plant*
The N concentration of individual plant organs is not constant. As the organ matures, its N concentration decreases as the protein is catabolized and translocated to younger tissues or to seeds (Dalling *et al.*, 1976) or even lost from the plant (Farquhar *et al.*, 1983). In particular, leaves and other vegetative tissues in annual plants are the main, if not the only, source of N to the growing seed. Depending on the demand for N by the seed, on the nature of the senescence process, and on the protein breakdown and transport system in the plant, the growth of the seeds can determine the longevity and photosynthetic capacity of the leaves (Sinclair and de Wit, 1976). Translocation also determines to a large degree the relationship between protein content and yield of grain (Kramer, 1979). The more efficient the translocation, the more rapidly do the leaves senesce, the shorter the 'green area duration'. Grain yield is then lower, but the N concentration of the grain can be higher.

3.5.1.6. *Nitrogen and canopy temperature*
Ample supply of N to the plant is necessary for unrestricted growth, especially in young plants. Under semi-arid conditions, where evaporative demand is high, vigorous growth involves greater transpiration (de Wit, 1958) and consequently more evaporative cooling of the plant canopy. Thus, a canopy deficient in N can be warmer than one well supplied with N, other conditions being equal (Seligman *et al.*, 1983). As a consequence, processes that are influenced by temperature are affected. Maintenance respiration requirements can increase and development rate can be speeded up so that plants deficient in N tend to ripen earlier than those growing on an abundant N supply. This can have important consequences for crop success or failure under semi-arid conditions where water availability during seed filling is a factor that limits yield, particularly in a mediterranean climate where transpiration demand also

increases as annual plants ripen their seeds. In some cases earlier maturity can save a crop.

3.5.2. *The simulation model PAPRAN*

3.5.2.1. *Short description of model additions*
The simulation model ARID CROP was extended to include the nitrogen balance in soil and vegetation, so as to study the sensitivity of production to changes in the nitrogen supply in semi-arid environments. This version was entitled PAPRAN, – Production of Annual Pastures limited by Rain and Nitrogen – and a detailed description is given by Seligman and van Keulen (1981). Here only its most salient features will be discussed.

The nitrogen balance in the soil is described on the basis of four different nitrogen pools: inorganic nitrogen, which includes ammonium, nitrate and any other inorganic components that may be present, organic nitrogen in fresh organic material, comprising mainly residues of last year's crop, organic nitrogen in the stable organic material, and nitrogen in the microbial biomass. The organic components are subject to microbial decomposition, which is described as a first order process, with a specific rate of decomposition dependent on the composition of the material being decomposed, and modified by soil temperature and soil moisture conditions. The rate of decomposition can further be limited by the maximum relative growth rate of the microbial biomass. The carbon of the decomposing material is partly used in microbial respiration and partly for growth of microbial tissue. As it is also assumed that the C/N ratio of microbial biomass is constant, the fate of nitrogen depends on the C/N ratio of the decomposing material: if it contains a relative excess of nitrogen, *i.e.* C/N smaller than that of the microbes divided by the growth efficiency (*i.e.* the fraction of the carbon in the organic substrate that is fixed in the microbial biomass), the net result will be mineralization of nitrogen, while in the opposite case the net result will be net immobilization of mineral nitrogen (Parnas, 1975).

Mineral nitrogen can volatilize from the upper soil compartment, it can be transported with water moving through the soil profile or it can be taken up by the plant roots. Denitrification is neglected in this description, as anaerobic conditions in the semi-arid zone seldom occur. This may well be an oversimplification as transient anaerobiosis in the rhizosphere can be quite common. Volatilization is described in a simplified way, as a first order process, depending on ammonium concentration, disregarding possible effects of soil pH or weather conditions. Transport through the profile is described according to the 'perfect mixing' principle, *i.e.* the rate of transport of nitrogen out of a soil compartment is calculated from the rate of water flow out of that compartment multiplied by a concentration obtained by assuming that all the mineral nitrogen in a compartment, as well as that transported over the upper boundary, are mixed with all the water in the compartment and that entering the compartment. Uptake of nitrogen by the root system depends on the

60

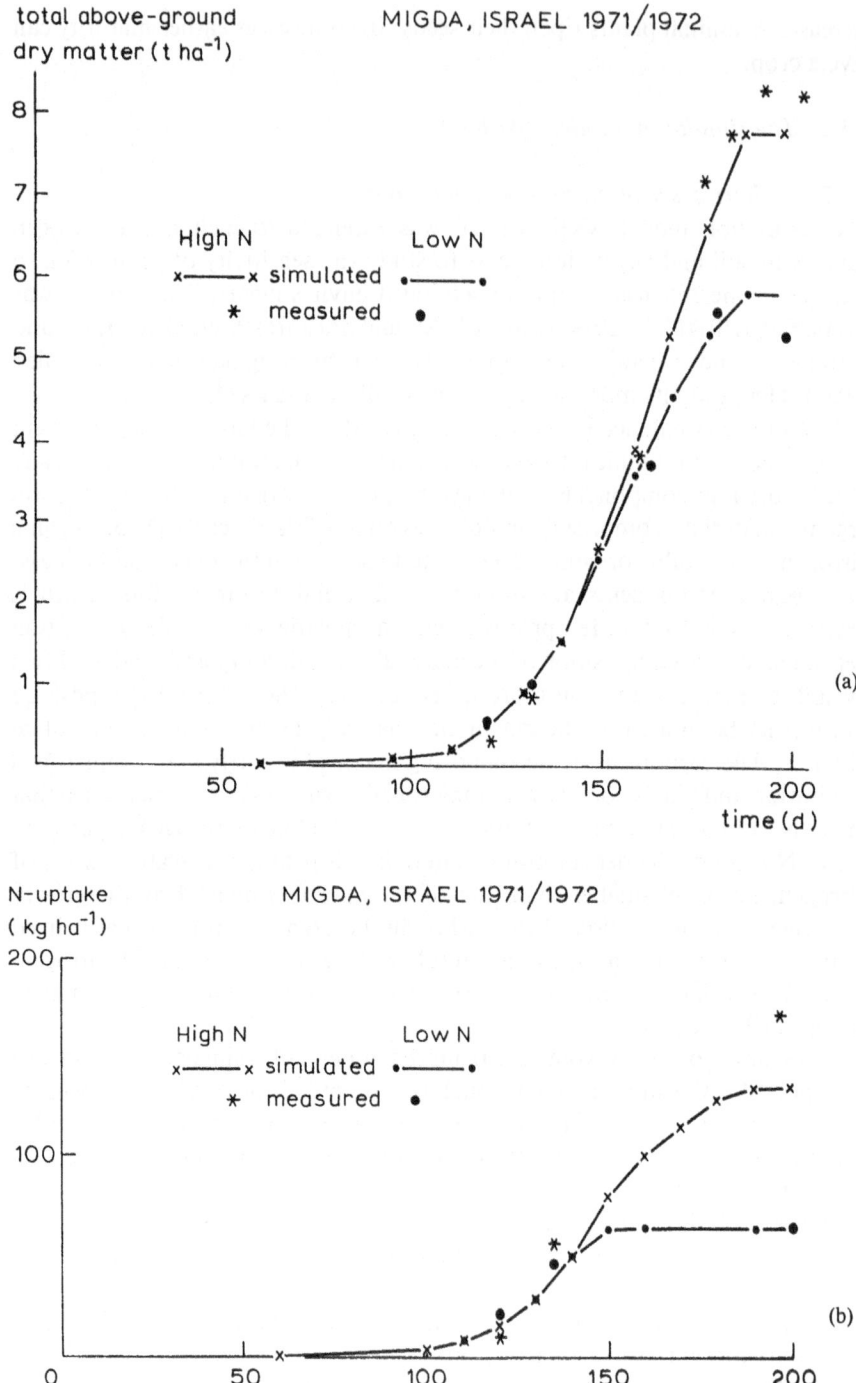

Fig. 3-13. Comparison of measured and simulated aboveground dry matter accumulation (a) and total nitrogen uptake (b) for spring wheat in Migda, Israel in 1971/1972 at two levels of N supply.

demand of the vegetation, the rate of transport with the transpiration stream and the possibility for transport by diffusion to the root surface. The potential uptake with the transpiration stream is calculated as root water uptake from each soil compartment multiplied by the concentration of nitrogen in that compartment. If this exceeds the demand of the crop, uptake is actively curtailed, to satisfy the demand exactly. If, on the other hand potential uptake by transpiration falls short of the demand, it is assumed that the remainder can be fully satisfied by diffusion, provided that sufficient mineral nitrogen is available in those parts of the rooted profile where soil moisture content is above wilting point.

The nitrogen demand of the vegetation is obtained as the sum of the demands of the various vegetative organs, which for each organ is calculated as its dry weight times a maximum nitrogen content, defined as a function of the development stage, minus the actual nitrogen content. If actual uptake falls short of the demand, nitrogen is distributed between the various organs in proportion to their respective demands. Nitrogen supply to the seeds is assumed to take place by translocation from the vegetative tissue only, and is regulated by temperature and the availability of amino acids derived from breakdown of proteins in these tissues. The potential N concentration decreases as the plant develops (van Burg, 1962), but if uptake cannot satisfy the demand, the nitrogen concentration in the various organs will drop below the potential value. The light-saturated assimilation rate of individual leaves is defined as a linear function of leaf nitrogen concentration (van Keulen and Seligman, 1987). If leaf nitrogen concentration falls below a development-stage dependent threshold value, leaf growth is hampered and the 'surplus' assimilates are distributed between roots and stems in proportion to their phenologically determined allocations of assimilate at the current development stage.

At still lower nitrogen concentrations, part of the vegetative mass dies, leaving a residual non-remobilizable nitrogen content in the dead tissue, the remainder being translocated to the living plant parts.

3.5.2.2. *Performance of the model*

The results of model calculations for the 1971/1972 season in Migda are given in Figure 3–13a for both high N, fertilized, and low N, non-fertilized, pasture. For the latter, the non-disked treatment as reported by van Keulen (1975) was chosen, because that represents the lowest nitrogen uptake. The simulated growth curve for the high N treatment closely follows the measured values for most of the season, but growth stops somewhat early, so that the simulated total dry matter production falls just short of the measured value. For the low N treatment the simulated growth curve is in good agreement with the measured values and the calculated final dry matter production is very close to the measured value.

Nitrogen uptake is illustrated in Figure 3–13b, which shows that for the high N treatment the simulated rate of nitrogen uptake is underestimated after day 120, resulting in a difference of almost 40 kg per ha at maturity. It is difficult

62

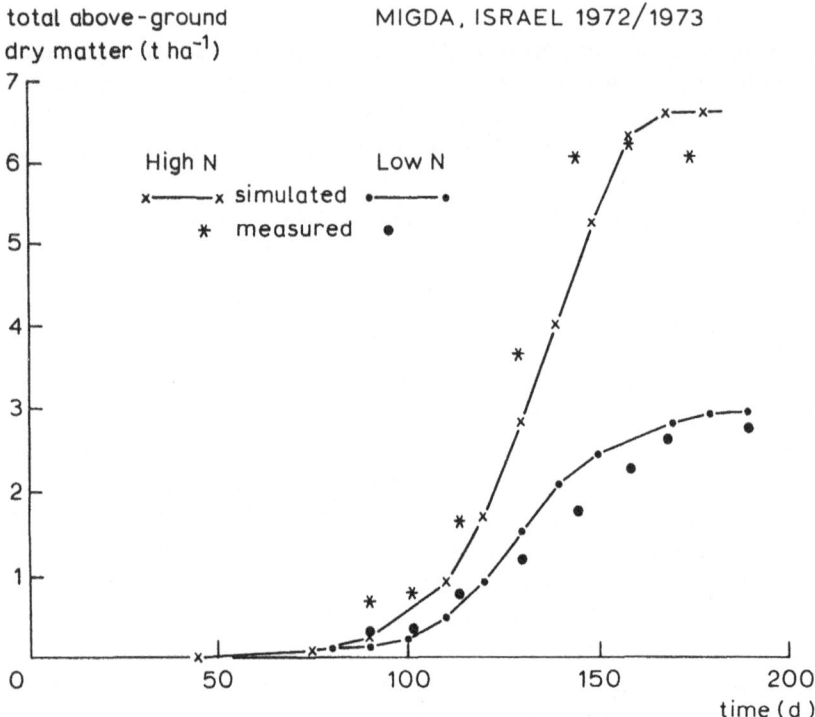

Fig. 3-14. Comparison of measured and simulated aboveground dry matter accumulation for spring wheat in Migda, Israel in 1972/1973 at two levels of N supply.

to explain this discrepancy. Higher values for nitrogen uptake can be simulated, but then peak biomass values result that are much higher than the measured values. A possible explanation is that the measured value is too high because of a sampling bias towards the end of the season when material sampled for chemical analysis excluded shedded leaves from which N was depleted. This sampling error can occur particularly in heavy canopies where loss of older, lower leaves can be considerable, but easily overlooked (Seligman *et al.*, 1976).

In Figure 3–14 the simulated growth curves for the '72/'73 season for high N and low N are presented. As the '71/'72 data were used for calibration of the model, these data represent the first independent validation data set. For the fertilized treatment, the simulated growth curve underestimates dry matter production slightly, throughout the season, but peak biomass is simulated with fair accuracy. For the low N treatment, measured and simulated dry matter accumulation are in close agreement for the whole season, individual points never deviating more than about 15 percent, which is within the accuracy of measurement. Experimental data for nitrogen accumulation are available only for the mature vegetation. In both the high N and low N treatments simulated N uptake is close to the measured values.

In the years 1965 through 1967, fertilizer experiments on native pasture were carried out at the Migda site in the framework of a comprehensive research

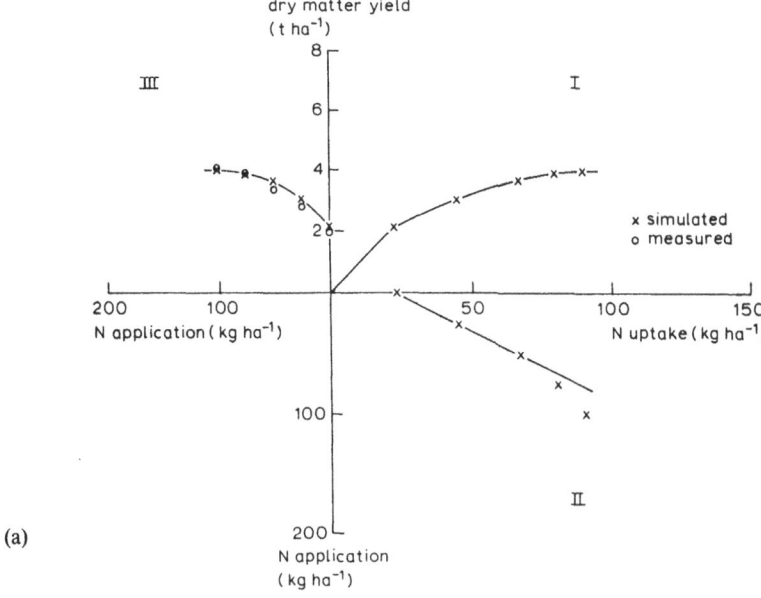

(a)

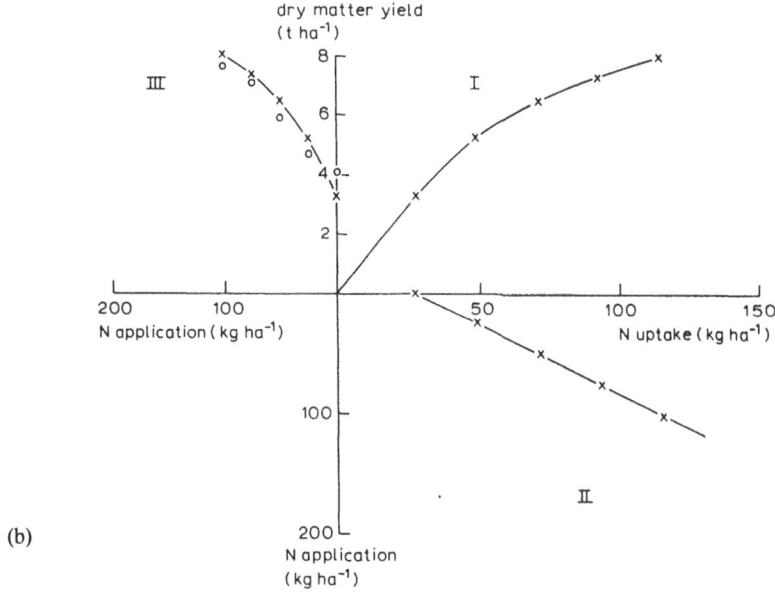

(b)

Fig. 3-15. The simulated relation between total nitrogen uptake and dry matter yield (I), that between nitrogen application and nitrogen uptake (II) and that between nitrogen application and dry matter yield (III) for natural vegetation in Migda, Israel for 1965/1966 (a), 1966/1967 (b) and 1967/1968 (c). The measured relation between nitrogen application and dry matter yield is given for comparison.

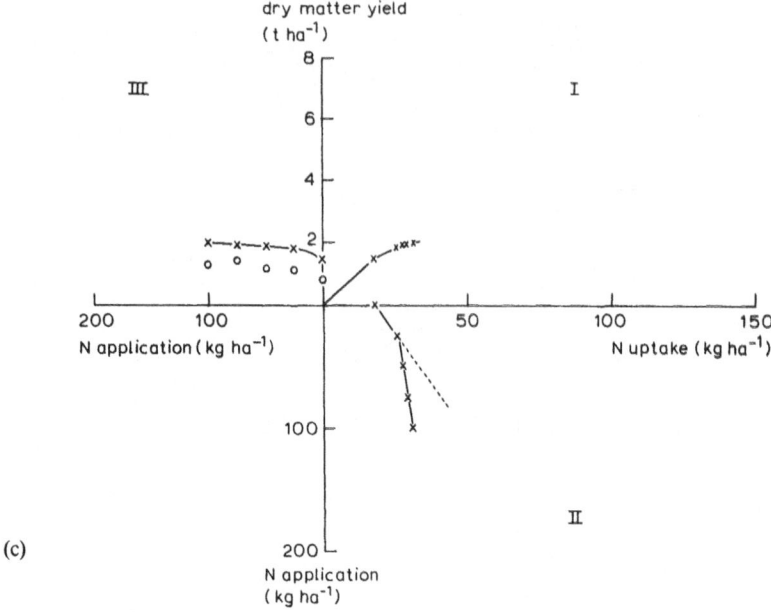

(c)

project on plant and animal production in the semi-arid region of Israel (Tadmor *et al.*, 1974, 1970, 1967, 1966). The plant material was not analysed for N concentration and only total dry matter yield at the end of the rainy season was recorded. For the 1965/66 growing season dry matter production is overestimated in the model for all nitrogen application rates (Fig. 3–15a). However, Tadmor *et al.* (1966) report, that "pasture yields at peak development were about 1.5 ton per ha dry matter", and it is not clear, why the control yield in the fertilizer experiment is reported as so much lower. The response to increased nitrogen availability is, however, very similar in both the experiment and the simulation. In this relatively dry season, total rainfall was 218 mm, but with a very unfavourable distribution, moisture availability determined growth for the greater part of the season, so that the additional available nitrogen was not taken up by the vegetation. This is illustrated by the relation between fertilizer application and crop uptake (Fig. 3–15a).

The '66/'67 growing season was a better season, with a total rainfall of 282 mm and a very favourable distribution. The response to increased N availability was positive up till the highest level applied (Fig. 3–15b). The application-uptake relation is a straight line, indicating a constant recovery fraction over the full range of application rates. This linear relationship has been detected in most fertilizer experiments involving nitrogenous fertilizer (van Keulen and van Heemst, 1979). The recovery fraction in this case is high at a value of 0.87, indicating that losses of nitrogen due to volatilization or leaching must have been negligible and that soil moisture conditions were favourable for uptake throughout the growing period.

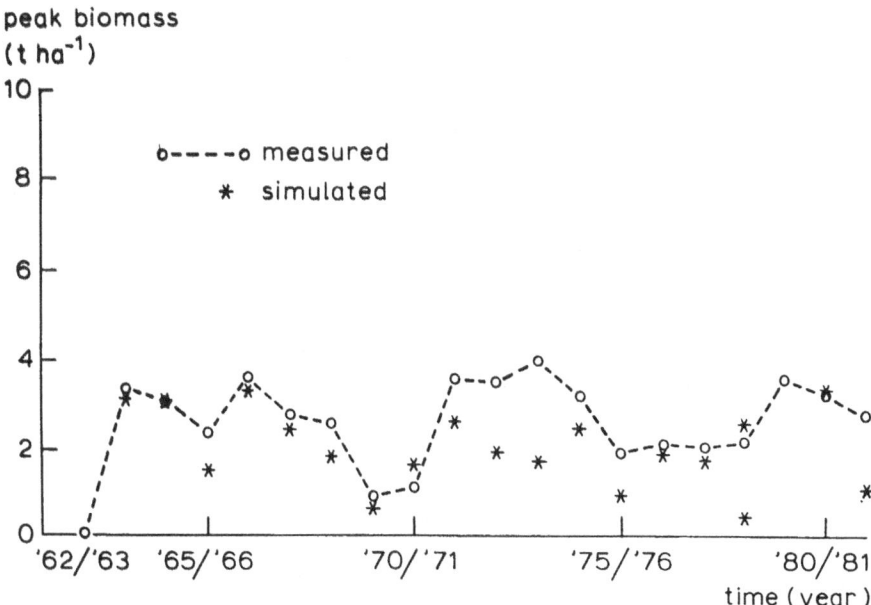

Fig. 3-16. Comparison of measured and simulated peak biomass of non-fertilized natural vegetation in Migda, Israel for a twenty-one year period.

The '67/'68 growing season was an intermediate case, with a total rainfall of 260 mm, but with a much less favourable distribution than in the previous season. Consequently, only 40 percent of the rainfall was used for transpiration at the highest nitrogen application rates (compared to two thirds in '66/'67). This is also reflected in the nitrogen response curve, that levels off at a total dry matter yield of about 4,000 kg per ha, associated with a nitrogen uptake of 90 kg per ha (Fig. 3–15c). The application-uptake relation also levels off at that value, indicating again that limited moisture availability was a constraint for higher nitrogen uptake. The recovery fraction for the lower application rates is identical to that for the preceding season. The simulated response curve is in close agreement with the experimentally determined curve.

The simulation model reproduced the growth response of natural vegetation to availability of water and nitrogen in a meaningful and recognizable way for three different seasons. This is a degree of validation that warrants the use of the model to examine long-term effects of fertilizer application, influence of soil properties, management practices, etc., on canopy growth, at least in the study area.

3.5.2.3. *Application of the model*
Simulation of long-term natural pasture production in the northern Negev region (Fig. 3–16) produced values for production similar to the observed values, albeit with substantial deviations in some years. One possible reason for these deviations, is that because of insufficient data, identical initial conditions

were assumed for each season, even though initial conditions can vary considerably from year to year. So, for instance, water shortage in one season may lead to higher nitrogen availability in the subsequent season (Harpaz, 1975). A ubiquitous problem is that experimental data can be very variable because of site and vegetation heterogeneity (Penning de Vries and Djitèye, 1982).

For the 20 years that were simulated, the average dry matter production with no added N was 1966 kg per ha, with a coefficient of variation of 51 percent (very similar to that under conditions where nutrients are not limiting). Average nitrogen uptake was 21.9 kg per ha, with a coefficient of variation of 28 percent. The variability in nitrogen uptake, under natural conditions is mainly related to differences in moisture availability that determine both the amount of N originating from natural sources (de Wit and Krul, 1982) and the ability of the crop to utilize nitrogen.

Over the 20-year period average transpiration as a fraction of annual precipitation was 0.21 (C.V. 33 percent). This value is about 50 percent lower than under conditions of optimum nutrient supply (Subsection 3.4.3.1), mainly due to the fact that with limited nutrients growth rates are lower, vegetative cover is less and evaporation losses are higher. Average water use efficiency related to shoot dry matter, is 7.25 (C.V. 41 percent) g per kg of precipitation, a value strikingly similar to those of other workers quoted in Subsection 3.4.1.2 above. This is substantially lower than in the situation where nutrients do not restrict growth and reflects a situation where nutrient availability, in this case nitrogen, can be a greater constraint to biomass production than water availability.

The model PAPRAN was also used to study the interaction of soil depth, water availability and nutrient supply in a mediterranean region (Seligman and van Keulen, 1989). The model was run with different maximum rooting depths, varying between 0.3 and 1.8 m, different annual amounts of rainfall, simulated by multiplying the measured rainfall at the Migda site by factors varying between 0.5 and 2, and different nitrogen application rates, varying between 0 and 150 kg per ha. The results, some of which are presented in Figure 3–17, show that soil depth can be a major determinant of biomass production in the strongly seasonal mediterranean environment. The shape of the iso-production lines in Figure 3–17a indicates, that in the dry range (132–264 mm precipitation annually), biomass production responded primarily to increased moisture availability, whereas the effect of nutrient availability was all but absent. In the intermediate range (annual precipitation between 264 and 396 mm), mean long-term biomass production responded to both increased moisture availability and to increased nitrogen availability. In the 'wet' range (396–527 mm precipitation), mean long-term production is determined almost solely by nitrogen availability, indicating that even though production varied between years with moisture availability, it was, as a rule, always limited by nitrogen deficiency.

The patterns of nitrogen uptake (Fig. 3–17b) differ from those of dry matter production, as uptake responded to increased nitrogen availability even in the

mean rainfall

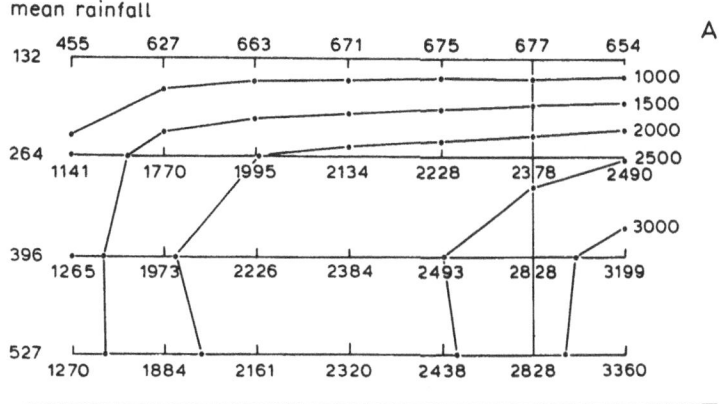

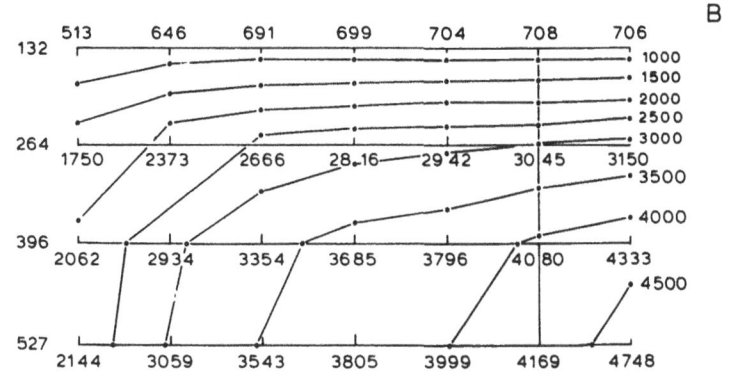

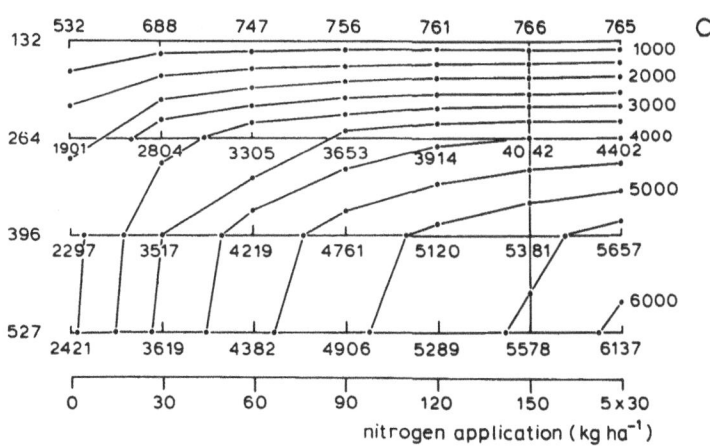

nitrogen application (kg ha^{-1})

(a)

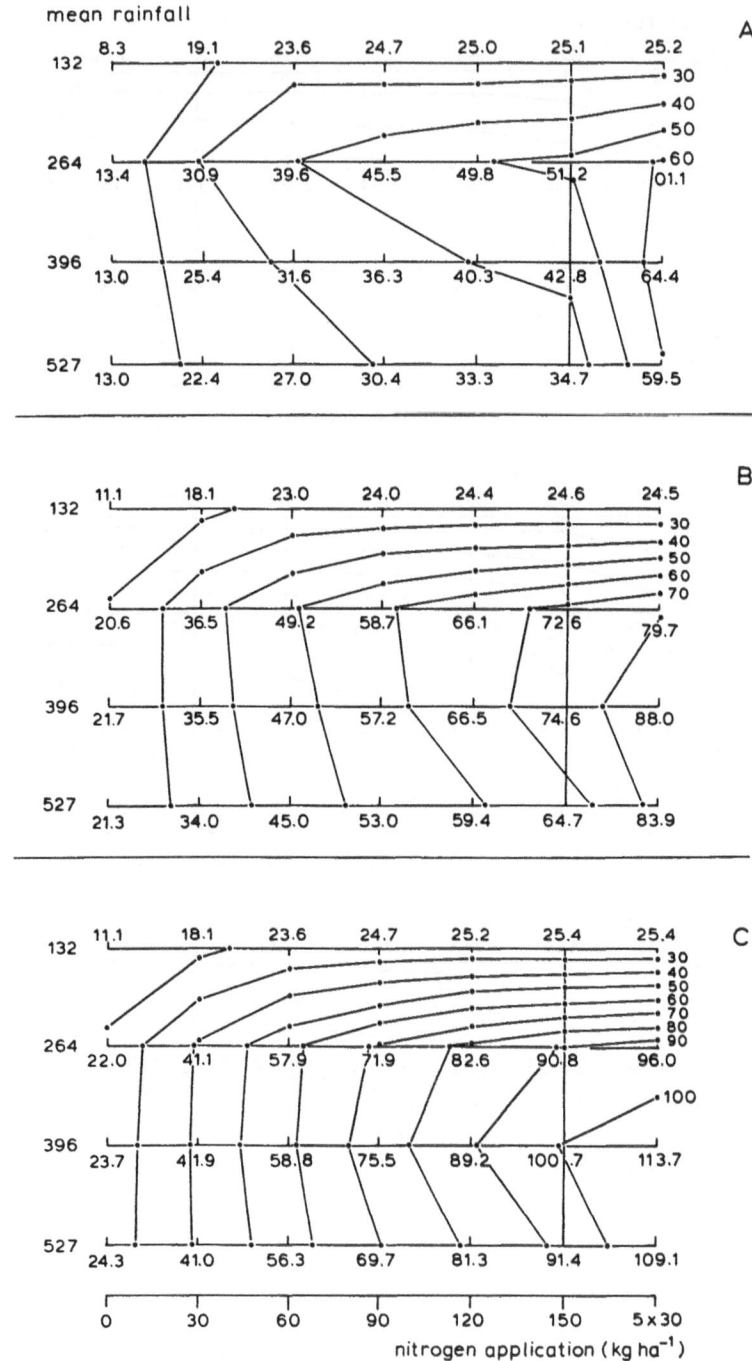

Fig. 3-17. Response of 21-year mean simulated total aboveground herbage production (a) and nitrogen uptake by the aboveground biomass (b) to rainfall and nitrogen application in three soil depth situations: (A) 0.3 m; (B) 0.6 m; (C) 1.5 m. Values of herbage dry matter production (a) and N-uptake (b) isolines are indicated at right in kg ha⁻¹ (Source: Seligman and van Keulen, 1989).

shallow soils, although it levels off earlier than in the deeper soils. There is clear evidence of decreasing nitrogen availability with increasing rainfall, iso-uptake patterns falling diagonally from left to right, already in the intermediate rainfall range. In the deeper soils this effect is less and occurs only at the higher fertilizer application rates. This phenomenon is caused by leaching losses, particularly in shallow soils. Such losses can be reduced by split application of the fertilizer: uptake is higher if it is assumed that nitrogen is applied in five doses, distributed over the first 150 days of the growing season.

3.6. Grain production in a semi-arid environment

In many semi-arid regions extensive animal husbandry based on sedentary systems or on different variations of nomadism, are integrated with the cultivation of small grains, especially in the more favourable parts of the region. Under suboptimum conditions that are common in these regions, the grain yield is not a constant fraction of the total dry matter yield. In order to study the implications of water or nutrient limitations on small grain yields, a crop model was developed for spring wheat, one of the most important crops in many semi-arid regions.

3.6.1. *Short description of the wheat model*

The model WHEAT (van Keulen and Seligman, 1987) is a simulation model based on the models ARID CROP and PAPRAN. The main features that distinguish it from the previous models are in the description of organ formation, leaf senescence and their dependence on both water and nitrogen limitation, grain growth and carbohydrate reserves.

3.6.1.1. *Growth of the crop canopy*
Germination starts after 'sowing', as soon as soil moisture content in the upper 10 cm of the soil profile is greater than 1.2 times wilting point. If these conditions persist uninterruptedly for seven days, germination is completed and the plants 'emerge'. If the soil dries out within four days after the onset of germination, the process is halted, but can resume after rewetting of the soil. If sufficiently dry conditions persist for more than six days, all the seeds die and resowing is necessary to establish a crop. The initial biomass at emergence is calculated from the sowing rate under the assumption that in the germination process about half the seed weight is used for respiration (Penning de Vries *et al.*, 1979), and that the remainder is equally divided between shoot and root. A large part of the germinating shoot is underground, so that only a small amount of the seed reserves contributes to initial photosynthetic tissue.

After emergence, growth and development are treated separately in the model. The order of appearance of the various plant organs is a crop characteristic and is not influenced by environmental conditions, but their rate of appearance is strongly influenced by temperature and daylength (van Dobben, 1962). In our model, photoperiodic effects are neglected and the rate of development, that is the inverse of the length of a given phenological phase, is defined as a linear function of average crop temperature. The relations used are different for the pre-anthesis and the post-anthesis phase, and relevant phenological events are defined separately for the two phases. The integrated value of the development rate yields the development stage, a unitless variable that assumes the value zero at emergence, 0.5 at anthesis and 1 at maturity, defined as the stage where the grain is dead-ripe.

After emergence, crop growth is dependent on the assimilates produced by the green plant parts in the photosynthetic process. Potential gross canopy assimilation is derived from the green area of the crop and incoming total short wave radiation. Leaf blades and sheaths, and, after ear emergence, stems and ears, all contribute to the total green photosynthetic area. The calculation of potential gross canopy photosynthesis is based on green area index, incoming radiation, latitude of the site and photosynthetic characteristics of the leaf (Goudriaan and van Laar, 1978).

Actual gross canopy assimilation is derived from the potential value by adjusting for the effect of water shortage, and carbohydrate reserve accumulation in the tissue. The effect of water shortage is described as in ARID CROP (Subsection 3.4.1.2). The effect of reserve accumulation on photosynthesis becomes operative when reserve concentration exceeds 25 percent on a total crop basis. Gross assimilation declines to zero as the concentration increases to 30 percent. This is only one of several possible representations of available evidence (Neales and Incoll, 1968).

Part of the assimilate fixed in the assimilatory process is used for maintenance respiration. The magnitude of losses by this process depends on the amount of material present, its nitrogen content and the prevailing crop temperature (Penning de Vries, 1975).

After satisfying the maintenance requirements the available assimilate is, until flowering, distributed among four sinks: leaf blades, stems (including leaf sheaths), roots, and a pool of reserve carbohydrates. The partitioning factors are defined as a function of the phenological development stage of the crop, and may be modified by crop water status and nitrogen status. The assimilates partitioned to roots, leaves and stems are converted into structural plant material, using a conversion efficiency that reflects the costs of growth respiration, defined as a function of the nitrogen content of the various organs (Penning de Vries, 1974). The reserve pool contains reduced sugars that may accumulate prior to anthesis and also under unfavourable growing conditions, such as water shortage or nitrogen deficiency. When conditions improve again, part of the reserves may be remobilized and used for growth of vegetative organs. When the vegetative organs have reached their final size, *i.e.* after full expansion of the flag leaf, and before grain growth starts, reserves continue to accumulate, mainly in the stem. After grain set, these reserves are used, together with current assimilates, for grain fill.

The growth rate of the grains is either determined by assimilate availability (source-determined) or by the potential growth rate of the grains (sink-determined) that is defined in the model as a function of grain number and prevailing canopy temperature.

The number of grains in the crop is the outcome of a long series of processes that determine the formation of the various precursor organs of the plant. These include the number of tillers, ears, spikelets per ear, florets per spikelet and finally the number of grains. In order to describe organ formation quantitatively, it is necessary to define a process that can account for the effects of growth and stress on organ formation. The algorithms used in the model are based on the hypothesis that the potential number of organs that can be formed by the crop is dependent on the rate of assimilate supply to the organs and the development rate of the crop. This assumes that an organ must reach a certain minimum size to be viable. The rate of organ formation is defined as the difference between the potential number that can be maintained at any moment by the assimilate flow and the current number, divided by a time constant for organ formation. The potential number is a linear function of assimilate flow and inversely proportional to the development rate. This description gives satisfactory results but depends on the parameter values that describe different cultivar characteristics, in particular the minimum amount of assimilate necessary to produce a viable organ. As these parameters have not been measured directly on plants they must be derived with the model for each cultivar. The temperature-dependent development rate is another cultivar characteristic.

Leaf area development of the crop is related directly to the rate of increase in dry weight of the leaf blades. Specific leaf area, *i.e.* the area per unit dry weight of the leaf blades is defined as a function of development stage as successive leaves become progressively thicker as the plant grows. Most experimental evidence indicates that leaf thickness in wheat is relatively invariable (Aase, 1976). Leaves have a limited life span, that can be expressed as a function of accumulated temperature (Gallagher, 1979; Ford and Thorne, 1975). In the model, leaves are distributed between age classes for each of which the temperature sum is tracked. If leaves die for any reason like water shortage, shading or nitrogen deficiency, the oldest leaves will be affected first. Surviving leaves will eventually die from senescence after the predetermined temperature sum has accumulated.

3.6.1.2. *Effects of nitrogen nutrition*
Length of the growing period
For most spring wheat cultivars the effects of daylength on development rate are small and generally insignificant. The driving force for development is the temperature of the stem apex, which can be approximated either by air temperature, or by canopy temperature. Nitrogen deficiency may result in stomatal closure as an adjustment controlled by internal CO_2 concentration (Radin and Ackerson, 1981; Bolton and Brown, 1980), or in reduced water

uptake by the roots due to increased root resistance (Radin and Boyer, 1982) and a consequent reduction in transpiration (Shimshi and Kafkafi, 1978). A lower rate of transpiration will change the energy balance of the canopy and hence its temperature. In the field, differences of up to 4 °C have been measured between crops optimally supplied with nitrogen and crops under nitrogen stress (Seligman *et al.*, 1983). Under these conditions, an indirect effect of nitrogen shortage on phenological development was observed, leading to shorter growth periods and earlier maturity. This may be the cause of 'emergency ripening' under N shortage. However, severe stress can delay or even stop development completely (Angus and Moncur, 1977).

Dry matter production
The effects of nitrogen status of the vegetation on dry matter production and yield are mediated through the basic processes of assimilation, respiration and partitioning. The rate of CO_2 assimilation at different levels of nitrogen content in leaves has been reported for many plant species. In all situations where the associated nitrogen content was also reported, a strong correlation existed between nitrogen content, either expressed on a weight basis or on an area basis, and photosynthetic performance. Analysis of a large data set by van Keulen and Seligman (1987) showed a linear relation between nitrogen content and light saturated assimilation rate. The correlation coefficients were very similar whether nitrogen content was expressed on a weight basis or on an area basis. The linear relation on a weight basis is incorporated in the model, with assimilation becoming zero at a nitrogen concentration of 0.0038 kg per kg and a slope of 7.25 kg CO_2 per ha per hour for each percent increase in nitrogen concentration in the leaves.

In general, no significant effect of leaf nitrogen content on initial light use efficiency has been shown (Cook and Evans, 1983a, 1983b; Wilson, 1975a, 1975b), but small differences in the slope could be responsible for measured differences in assimilation rate between canopies with different leaf nitrogen contents. However, due to uncertainty, this has not been incorporated in the model.

Partitioning of assimilates
The assimilates fixed in the photosynthetic process are used for respiration and for growth of the various plant organs. Sink strength, which probably is related to the number of growing cells in a particular organ (Sunderland, 1960), is an important determinant of assimilate distribution at any moment.

Maintenance respiration, providing energy for biological functioning, presumably has first priority. The relative rate of maintenance respiration, *i.e.* the energy requirement per unit of dry weight, is directly related to the nitrogen content of the material, possibly as a result of the higher levels of metabolic activity in plants that grow more vigorously because of high nitrogen concentrations in the vegetative tissue (Penning de Vries, 1975). In the model this is accounted for by a multiplication factor that ranges between 1 and 2, over the range of 'absolute' minimum to 'absolute' maximum nitrogen content, which is about the range found between nitrogen-rich and nitrogen-poor wheat crops (Pearman *et al.*, 1981).

The assimilate remaining after satisfying the maintenance requirements, is available for production of structural plant material and is allocated to five sinks: leaf blades, stems and leaf sheaths, roots, grains and a pool of reserve carbohydrates. Under optimum growing conditions partitioning is a function of the phenological state of the crop only, representing the varying sink strength of the various organs. Under sub-optimum growing conditions partitioning changes (*c.f.* Brouwer, 1962). Nitrogen shortage in the vegetation favours growth of the roots relative to the above-ground material (Cook and Evans, 1983a), according to the functional balance principle of Brouwer (1963). The partitioning between leaf blades and other above-ground organs also changes in nitrogen-deficient conditions and generally results in lower leaf weight ratios (van Os, 1967; McNeal *et al.*, 1966). These results, however, do not permit derivation of the instantaneous effect of nitrogen deficiency on assimilate partitioning. In the model, nitrogen stress at any point in time is defined as the difference between the maximum nitrogen content of any organ at the appropriate development stage and the actual nitrogen content, expressed as a fraction of the difference between the maximum and the minimum nitrogen content. The effect on partitioning is described schematically by assuming a growth check on the leaf blades whenever nitrogen stress reaches a certain threshold value. The resulting 'surplus' carbohydrate is partitioned among roots, stems and sheaths, and the reserve pool, in proportion to their phenologically determined allocations of assimilate at the moment of stress.

Conversion of assimilate into dry matter
The assimilate allocated to the various sinks consists of a mixture of carbohydrate and nitrogenous compounds, that are converted into structural plant material. The amount of energy consumed in the growth process is termed 'growth respiration', and its magnitude depends on the chemical composition of the material being formed (Penning de Vries, 1974). The rate of increase in dry weight of an organ is thus obtained by dividing the rate of assimilate supply by the specific assimilate requirement factor, defined as 1.2 times the fraction carbohydrate in the material currently formed plus 2.27 times the fraction of protein (Penning de Vries, 1974). This description results in higher 'conversion efficiencies' for tissues with lower protein content.

Transpiration
Many studies have shown that water use efficiency increases with higher nitrogen availability (Section 3.3.2; Black, 1966; Viets, 1962). This may partly be explained by the fact that plants growing under nutrient stress generally have a smaller leaf area, so that canopy closure occurs later. As a consequence, a larger proportion of water is lost directly through soil surface evaporation, which reduces water use efficiency (Section 3.3.2). The effects on transpiration

efficiency are less conclusive (Subchapter 3.1; Bolton and Brown, 1980; Goudriaan and van Keulen, 1979; Wong *et al.*, 1979).

In a comprehensive study on the interactions between water and nitrogen stress, conducted by Radin and associates (Radin, 1983; Radin and Boyer, 1982; Radin and Ackerson, 1981; Radin and Parker, 1979a, 1979b) it was found that in nitrogen-deficient plants, stomatal closure occurs at much higher plant water potentials than in plants adequately supplied with nitrogen. In addition, evidence for greater stomatal opening with better nitrogen nutrition has been found in rice (Ishihara *et al.*, 1978; Yoshida and Coronel, 1976) and wheat (Shimshi and Kafkafi, 1978).

In the model a linear relation is assumed between stomatal conductance and nitrogen content in the leaf blades (Yoshida and Coronel, 1976). This is based on the assumption that a constant ratio between external and internal CO_2 concentration is maintained and that any decrease in assimilation rate will lead to proportional stomatal closure and decreased transpiration.

Organ formation
The effect of nitrogen status of the canopy on leaf area expansion and on organ formation is mainly through the effect on assimilate supply. However, the rate of leaf area expansion has been shown to be linearly related to nitrogen content (Greenwood, 1966; Greenwood and Titmanis, 1966). Tiller formation is also directly affected by the nitrogen status, very much in analogy to leaf expansion (Yoshida and Hayakawa, 1970; Aspinall, 1961). Therefore, in the model leaf area expansion and tiller formation are also related to the relative nitrogen content in the leaf blades, on a scale running from the content observed in severely depleted leaves to a phenology-dependent maximum value.

Translocation of nitrogen to the grain
The grains receive most of their nitrogen in reduced form, generally as amino acids from the vegetative parts of the plant (Nair *et al.*, 1978). The rate of nitrogen accumulation in the grains appears to be constant during a considerable part of the grain filling phase (Vos, 1981; Sofield *et al.*, 1977). The rate of accumulation at any moment may be limited by the potential rate of accumulation (sink) or by the rate of supply from the vegetative parts (source).

The rate of nitrogen depletion from the vegetative parts is fairly constant till its nitrogen content approaches a value of about 0.01 kg per kg (Dalling *et al.*, 1976). Such a constant rate over a wide range of nitrogen contents can be explained as withdrawal from a more or less constant pool of amino acids (Hanson and Hitz, 1983).

3.6.2. *Performance of the model*

The calibration and validation of the model have been described elsewhere (Van Keulen and Seligman, 1987). An example of model performance under different conditions is a run on data from two successive growing seasons at Tel

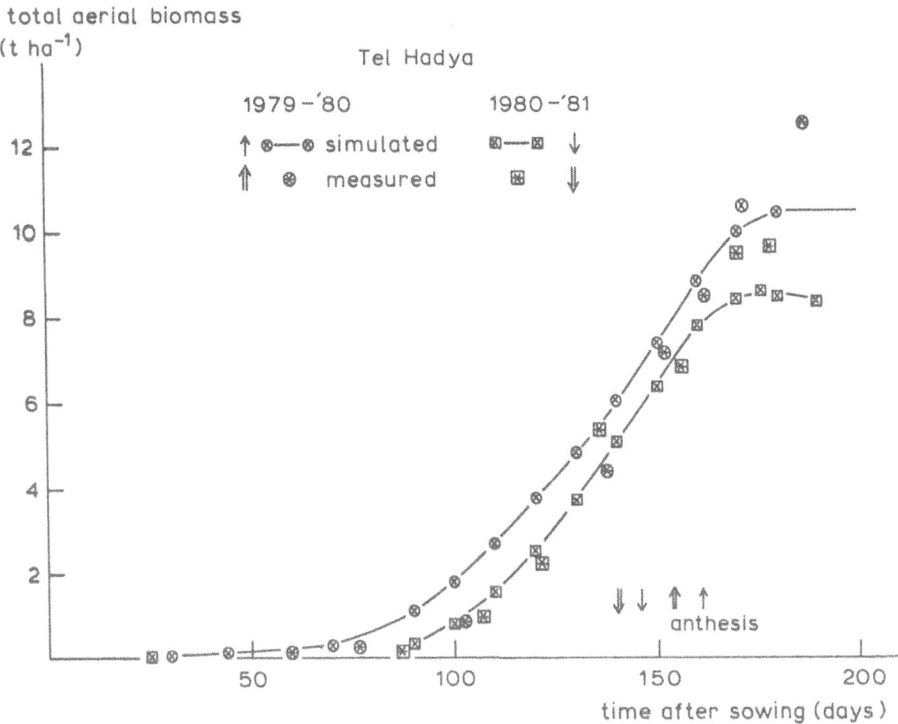

Fig. 3-18. Measured and simulated aboveground dry matter accumulation for spring wheat in Tel Hadya, Syria for the seasons 1979/1980 and 1980/1981.

Hadya, Syria, (Fig. 3–18) one of the experimental sites of the International Center for Agricultural Research in the Dry Areas (ICARDA), as reported by Stapper (1984). In the 1979/80 season dry matter accumulation was slightly overestimated by the model in the early growth stages, and underestimated towards the end of the growth cycle. The high growth rate measured in the field towards maturity looks suspicious, given the prevailing soil moisture conditions, but there is no obvious reason to doubt the data, except for the fact that very often experimental data are strongly influenced by heterogeneity in the field (van Keulen and Seligman, 1987; Biggar, 1984). Grain yield and yield components are predicted by the model with reasonable accuracy (Table 3–7). Rainfall in the 1980/81 growing season was lower than in 1979/80 and hence total dry matter production was lower, both observed and in the model, which predicted the growth curve more accurately than in the previous season. Grain yield was predicted very accurately, but the simulated yield components deviated substantially from reality for reasons that are not clear.

Extensive evaluation of model behaviour by van Keulen and Seligman (1987) showed that even though the model was not able to reproduce reality in all cases with sufficient accuracy, it reproduced recognizable and reasonable wheat crop

Table 3-7.
Comparison of measured and simulated yield components for two growing seasons at Tel Hadya, Syria.

Yield component	1979/80		1980/81	
	Observed	Simulated	Observed	Simulated
Ear number (ears m^{-2})	359	345	291	261
Grain number (grains ear^{-1})	42.9	48	41	32
Grain weight (10^{-6} kg $grain^{-1}$)	27.3	25	28.7	40
Grain yield (kg ha^{-1})	4,204	4,227	3,420	3,352

behaviour patterns. It seems that it can be used with a fair amount of confidence for a large group of sensitivity analyses and for analysis of the effects of climatic variability, cultivar characteristics or agronomic practice on growth and yield of wheat grown in semi-arid conditions.

3.6.3. *Application of the model*

3.6.3.1. *Long-term variability*

The performance of a wheat crop under rainfed conditions in a semi-arid region over a relatively long period was simulated using 21 years of meteorological data from the Migda site. The spring wheat cultivar used was the 'optimum' cultivar described by van Keulen and Seligman (1987). The agronomic conditions for the 'standard' run assumed a fixed sowing date (November 29) with no resowing after crop failure; abundant nitrogen fertilizer (150 kg per ha); two-fifths as a pre-sowing dressing and the remainder in three equal doses, 30, 60 and 90 days after sowing. (This practice may not necessarily give the highest grain yields, as strong interactions exist between nitrogen supply and moisture availability (Amir *et al.*, 1982)). In relatively unfavourable rainfall years, abundant nitrogen may favour luxurious vegetative growth with high water use, resulting in earlier water depletion and lower grain yield).

The results in Figure 3-19 show that for the 21-year period, in which annual precipitation varied between 78 and 414 mm, the inter-annual variability was extremely high. Grain yields varied between 0 (1962/63) and 4,650 (1982/83) kg per ha on a dry weight basis. Average grain yield for the period was 2,080 kg per ha with a standard deviation of 1,806 kg per ha. Total dry matter production averaged 6,826 kg per ha, with a standard deviation of 4,018 kg per ha. The harvest index varied between 0 and 0.41, with an average value of 0.24 and a standard deviation of 0.13. The average transpiration efficiency expressed in g dry matter produced per kg water transpired is 3.44, and although there was considerable variation from year to year, it was one of the more stable characteristics with a standard deviation of 0.8. The water use efficiency, *i.e.* g

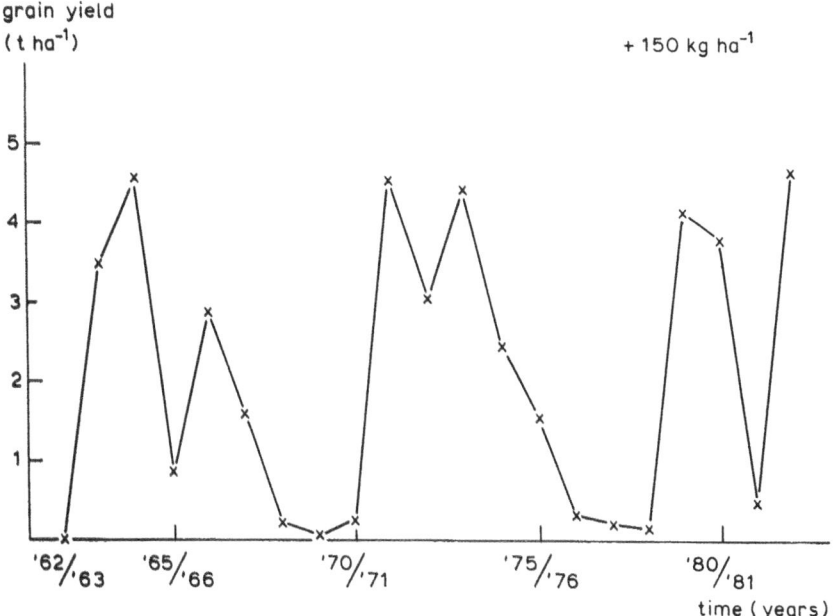

Fig. 3-19. Simulated grain yield of spring wheat, fertilized with 150 kg N per ha, for a twenty-one year period in Migda, Israel.

dry matter produced per kg total moisture input by rain was on average 2.35, with a standard deviation of 1.05. The average grain yield per unit rainfall was 0.96 g per kg of water, or 10 kg per mm per ha.

The results indicate that the between season variability in most crop characteristics was much higher than that in total dry matter production and in annual precipitation. The main reason is that the sensitivity of grain numbers and yield to water shortage is added to the sensitivity of total dry matter production. The increase in sensitivity is almost 50 percent: the coefficient of variation is 87 percent for grain yield compared to 60 percent for total dry matter production and 45 percent for water use efficiency compared to 22 percent for transpiration efficiency.

When these results are compared to those for the natural vegetation over essentially the same period (Subsection 3.4.3.1), average total aboveground dry matter production for the wheat is higher by about 25 percent, and transpiration efficiency related to aboveground dry matter about 40 percent lower. Both results are related to the fact that the wheat crop has a longer growing season than the natural vegetation. The lower transpiration efficiency is due to the fact that environmental conditions, *i.e.* radiation, temperature and humidity in April and May create much higher evaporative demands than in the preceding period. Phenological development of the natural vegetation is such that by about April 1 the growth cycle is completed. The wheat crop matures well into May and in favourable rainfall years dry matter accumulation continues till the middle of that month (van Keulen and Seligman, 1987; van

78

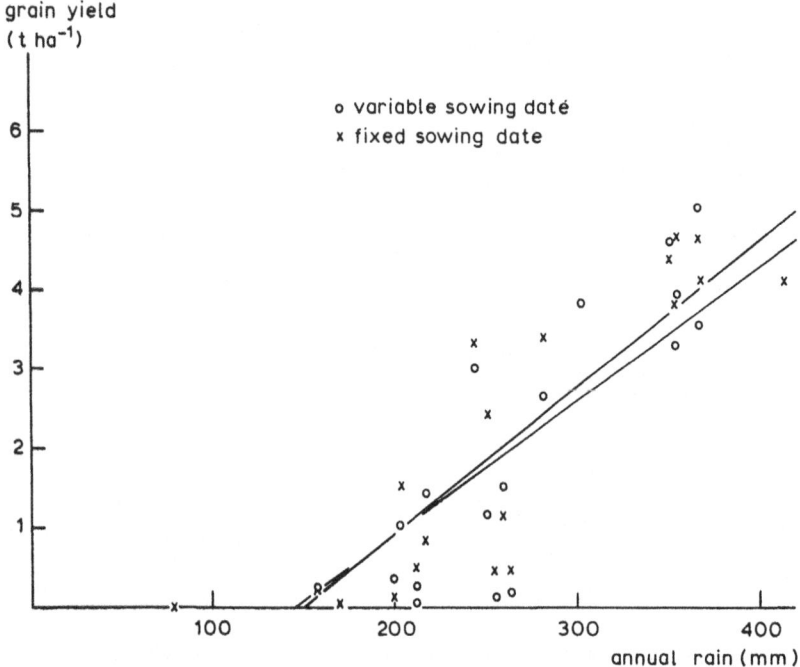

Fig. 3-20. The relation between annual rainfall and grain yield of spring wheat for a twenty-one year period in Migda, Israel for a fixed sowing date (November 15) and for a sowing date dependent on rainfall.

Keulen, 1975). Soil depth had a much larger effect on wheat performance than on pasture growth. Reducing the soil depth from 180 to 60 cm reduced average aboveground dry matter production of the natural vegetation by only 10 percent (Subsection 3.4.3.2), whereas in the wheat crop the decline was almost 25 percent and average grain yield was only half at 1018 kg per ha with a standard deviation of 1025 kg per ha. The lower dry matter production expresses itself almost entirely during the last part of the growing cycle, *i.e.* during the grain filling period.

3.6.3.2. *Effects of management strategies*
Sowing strategy
In calculating long-term variability in wheat yields, a fixed sowing date was assumed, irrespective of weather conditions. In view of the erratic nature of rainfall in semi-arid regions, often a more opportunistic strategy is followed, and the crop is sown only after the first 'effective' rains have fallen. In some developing regions that is almost a necessity, because the soils are so hard when dry that it is impossible to work them with animal-drawn implements. In those situations seedbed preparation, and consequently sowing, has to wait until the soil is wet.

   To examine the effects of an opportunistic strategy, the 21 years were re-run, with sowing date determined by soil moisture status. The wheat is 'sown' only after the water available for plant growth exceeds 30 mm in the top 30 cm of the profile. (Total available moisture in this soil equals 1.55 mm per cm.)

The results in Figure 3–20 show that there is very little difference between the yields for the fixed sowing date (in this case November 29) and the variable sowing date, although a slight tendency exists for somewhat lower yields for the fixed sowing date in low rainfall years and the reverse in high rainfall years. That would suggest that opportunistic behaviour, as practiced by some of the Bedouin in the region, is mainly a matter of costs of soil preparation which can be somewhat lower in moist soil than in dry soil. For the 21-year period, for which average precipitation was approximately equal to the long-term average, average grain yield was 1,947 kg per ha, with a standard deviation of 1,764 kg per ha. That is slightly lower than the 2,080 kg per ha calculated for the fixed sowing date. Average total dry matter was 6,477 ± 4,102 kg per ha, compared to 6,826 kg per ha for the fixed sowing date. All other characteristics of the crop, such as harvest index and transpiration efficiency are very similar to those reported in the preceding subchapter.

Fertilizer strategy

The response of wheat yields to fertilizer application in two contrasting seasons is given in Figure 3–21. The relation between grain yield and total uptake of nitrogen by the shoot is given in the upper right hand side of the figure and the relation between fertilizer application and nitrogen uptake in the lower right hand side (van Keulen, 1977; de Wit, 1953). In the 1964/65 season, rainfall was 414 mm, well-distributed over the season, and grain yield showed a positive response up to an uptake of 120 kg N per ha. Higher uptake caused a yield depression, mainly because of higher respiration requirements of nitrogen-rich material, resulting in lower availability of carbohydrates for grain fill. In the 1965/66 season, with a rainfall of 218 mm but a very unfavourable distribution, there was a negative response beyond an uptake of about 35 kg per ha. Here the major reason is that increased N availability early in the season stimulates vegetative growth and hence water consumption, so that water is depleted completely by the time grain fill starts. Similar responses to fertilizer application have been recorded for natural vegetation at the Migda site in these two seasons (Tadmor *et al.*, 1966; Ofer *et al.*, 1967).

The relation between fertilizer application rate and uptake is linear, for the 1964/65 season over the full range of applications used, for 1965/66 up to an application rate of 60 kg per ha. This linearity stems from the fact that the processes determining availability of nitrogen to the vegetation were modelled as first order equations. Within the range of practical applications this approach is supported by many experimental data (van Keulen, 1977). The slope of the line with respect to the vertical represents the efficiency of fertilizer uptake or the recovery fraction. For both seasons the recovery fraction is high with values of about 0.95, *i.e.* practically all the applied fertilizer, assumed to have been given as a basic dressing at sowing, is recovered in the crop. In both

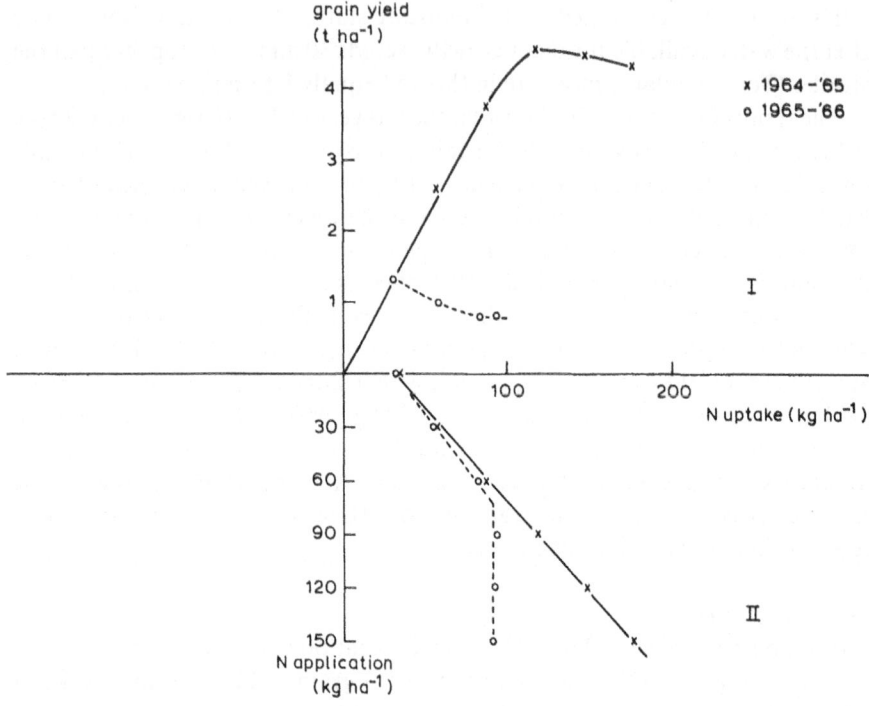

Fig. 3-21. The simulated relation between grain yield and nitrogen uptake (I) and that between nitrogen application and nitrogen uptake (II) for spring wheat in Migda, Israel for 1964/1965 and 1965/1966.

seasons rain started relatively shortly after sowing, so that losses by volatilization were probably small. Losses through denitrification were assumed to be low, but losses through leaching are accounted for in the model. In the 1965/66 season, uptake levels off above an application of 60 kg per ha; at that point the vegetation is 'saturated' with nitrogen throughout the growth cycle and uptake is actively prevented.

These results illustrate that fertilizer application in the semi-arid zone should be done judiciously, preferably in such a way that the application rate is related in some way to the availability of moisture to the crop.

3.7. Conclusions

It has been shown in this chapter, that models describing the processes that determine moisture availability to the vegetation and the impact of temporary water shortage on crop performance are quite realistic and can account for the main attributes of crop growth under semi-arid conditions. These models can be used to study water-limited production, especially in regions where insufficient experimental information is available. It has also been shown, however, that in each new situation the important special phenomena should first be identified so that suitable site-specific information on soil conditions and plant properties can be included in the model.

The relatively small amounts of rainfall and its highly variable inter- and intra-seasonal distribution are so characteristic of the semi-arid region, that water availability is commonly assumed to be the dominant factor that limits plant production. However, it has been shown that despite the low amounts of rainfall it is not at all unusual for crop and pasture growth to be constrained by nutrient deficiency. Production can be increased by amelioration of the deficiency with chemical fertilizers or, in some cases, by introduction of N-fixing leguminous species. In practice, economic considerations will determine what technological solution can be implemented.

The variability in marketable yield of grain crops is even greater than the variability in total dry matter production, because the grains are filled towards the end of the growing season, when the variability in rainfall is even higher than during the main rainy period. Increasing nutrient availability for wheat is not always a good solution because the resultant luxurious vegetative growth with its associated transpiration may so deplete the available soil water, that insufficient water is available during the critical grain filling period. When that occurs, grain yield can be lower with added N than without. The implications for management are treated in Chapter 5.

4. Sheep husbandry for lamb production in a semi-arid mediterranean environment

R.W. BENJAMIN

4.1. The grazing environment

In the mediterranean-type arid zones, rainfall is strongly seasonal with large interannual fluctuations. During the growing season pastures are of high quality with a digestibility and a crude protein content of up to 80 and 20 percent, respectively (Tadmor et al., 1974). With senescence and eventual death, pasture quality falls to digestibility levels and crude protein contents of at best 50 and 7 percent, respectively (de Ridder et al., 1986). The high quality pastures can support productive functions of lactation and lamb growth, but the low quality dry pastures can, at best, support sheep maintenance. The management of the animal production process must adjust to the continually changing balance between the animal feed requirements and their feed source.

In dry mediterranean regions, small domesticated ruminants, particularly sheep and goats, are better adapted than cattle for successful animal production because of their shorter biological life cycle and faster turnover of productive functions: gestation is about 155 days; lambs can be weaned after 45–60 days and be sold at up to 6 months of age; replacement offspring can reach sexual maturity in 6 to 12 months, depending on breed and feed supply. The productive functions of sheep and goats more closely follow the seasonality of annual herbage availability and quality. This is illustrated in Figure 4–1. Sheep, because of their mouth and muzzle characteristics (Mason, 1951), are able to graze closer to the ground than cattle by either biting low standing plants or gathering plant residues with their lips. This ability allows them to be more selective in their grazing than cattle, and to collect their feed requirements more easily at lower herbage availabilities. In addition, sheep are more fertile than cattle and some breeds (e.g. Dorset Horn) are even able to lamb twins twice a year. This makes sheep husbandry systems flexible and able to adapt quickly to changing environmental conditions and to market demands. Within-season adaptability is also possible: when pasture production is low, nutritional requirements for lactation can be significantly reduced by shortening the length of lactation to a minimum of 45 days by early weaning of lambs onto solid feeds. When pasture production is high, out-of-season lambing can utilize surplus herbage availability.

Small ruminants have been the predominant grazing animals in the semi-arid

Th. Alberda et al. (eds.), Food from Dry Lands, 83–100.
© 1992 *Kluwer Academic Publishers*.

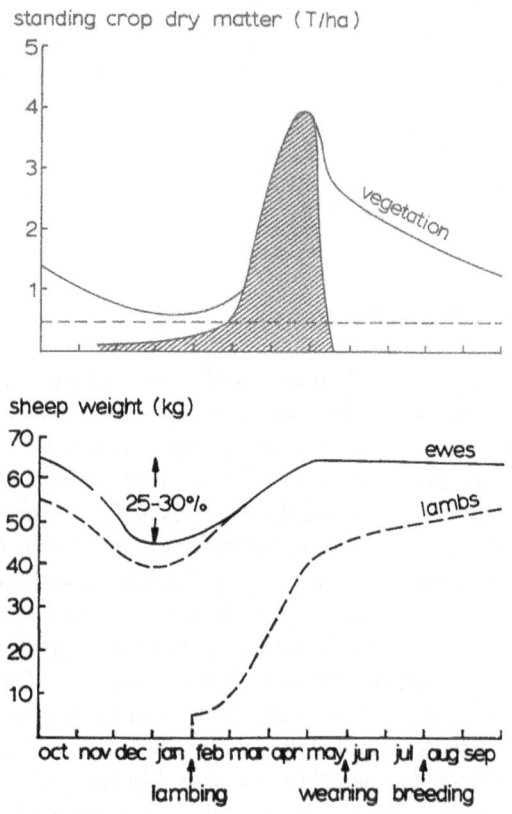

Fig. 4-1. The seasonal development of natural pasture and the liveweight changes of grazing ewes and their lambs in a 250 mm rainfall year (Source: Tadmor *et al.*, 1974). Upper part: Seasonal development of annual pasture in Migda; striped area denotes green forage; during the rest of the year the vegetation is stem-cured hay. Lower part: Ewe weight fluctuations and lamb development adapted to the forage seasonality. Both parts are semi-schematic abstractions, as extreme variation occurs between years.

mediterranean countries since biblical times and have been a source of milk, meat, wool as well as a capital investment (Mason, 1967; Epstein, 1970). Traditionally, the people in the region prefer fresh mutton and lamb to other meat products. In the Middle East the demand for meat has increased with the increase in urban population and the greater purchasing power in the Gulf States. This increasing demand for small ruminant products opens up possibilities for more intensive animal production in the region. In this chapter some of the more important factors that determine sheep performance in the semi-arid mediterranean context are discussed. As in preceding chapters, the situation will be described in more detail for an area in the northern Negev desert in Israel, where the rains fall between October and May and vary between 40 and 400 mm per year.

4.2. Breeds of sheep in Israel

The Awassi breed of sheep is common in much of the Middle East and appears to have evolved under the nomadic and transhumant way of life practiced traditionally in the region from biblical times until about 1950 (Epstein, 1971). It is suited to extensive systems of management with few inputs, apart from shepherding and some compulsory vaccinations. The production per animal is low. The breed is characterized by long legs, a non-mutton type body, a fat tail and a mature liveweight of approximately 50 kg. Anoestrus lasts from February to May (Amir and Volcani, 1965), fertility is one lamb per lambing and it responds to low levels of nutrition during pregnancy and lactation following lambing by not conceiving in the following breeding season. It can have a high initial milk production of up to two litres per day that rapidly falls, so that per annum production is rarely above 100 litres (Finci, 1957). Being native to the region, it is resistant to all but the most virulent of contagious diseases such as foot-and-mouth disease and contagious abortion.

Under nomadic conditions milk was produced for home consumption and mutton for sale to urban centres in order to buy clothing and other essential commodities. Essentially, it was a subsistence system. After 1950, intensification of production was motivated by an increasing urban market for animal products, including lamb and sheep milk products. This created higher prices and a more favourable price ratio between animal products and supplementary feeds, particularly concentrates. At the same time, the Awassi breed was improved in Israel by selective breeding for high milk production, so that at present the improved ewe has a mean mature liveweight of 60 kg and can produce a mean of 350 litres of milk per lactation. This is the minimum for registration in the Sheep Breeders' Association Stud Book (Finci, 1957). To improve the fertility of the breed, while maintaining its improved milk production, the Awassi was crossed with imported German East-Friesian rams and the resulting cross (the Assaf breed) produced 1.5 lambs per lambing (Goot, 1986). From 1970 onwards, overproduction of milk and higher labour costs led to intensive lamb production systems, using imported German Mutton Merino sheep. This breed is characterized by a mature liveweight of 65 kg, relatively short legs, a mutton type body and a fine wool fleece. The ewe does not have a well-defined anoestrus period and reaches a mean fertility of up to 1.5 lambs per lambing (Morag et al., 1973). Today the breed is used in intensively managed systems to lamb three times every two years, achieved by the use of hormones for out-of-season breeding. This intensification of lamb production, combined with the breed's rather poor milk production, has resulted in little or no grazing, heavy reliance on supplementary feeding, early weaning of lambs and high capital costs in buildings and equipment.

The high costs of production associated with these systems necessitated the use of a breed with a higher prolificacy. From 1979 Finnish Landrace rams were introduced to cross with both German Mutton Merino and Awassi ewes to produce Finn crossbred ewes with a potential fertility of 1.8 lambs per lambing

(Goot *et al.*, 1976), no well defined anoestrus period and potentially able to lamb three times every two years. The Finn x Awassi cross in particular, by having a higher milk production, has proved to be suitable for intensified lamb production in grazing systems as less lambs need to be raised by artificial rearers. However, like the German Mutton Merino, the use of Finn crosses for intensified lamb production has been mainly limited to a feedlot feeding system.

4.3. Reproduction and breeding practice

4.3.1. *Definition of terms*

Fertility is the ability of the ewe to conceive and lamb and is expressed as the number of lambs born per ewe per lambing; *net fertility* is the number of live lambs at weaning per ewe per lambing; *prolificacy* is the mean number of lambs born per ewe in the flock per year; *net prolificacy* is the number of lambs weaned per ewe in the flock per year. In this context, 'ewe' refers to all reproductive females, including hoggets 6–18 months old. *In-season breeding* refers to the natural breeding season for the breeds native to the region, normally from July to December. This leads to *in-season lambing* from November to April. *Out-of-season breeding* refers to breeding when the sheep native to the region are in anoestrus, in Israel from January to May; *out-of-season lambing* then occurs generally from June to September.

4.3.2. *The breeding season and sheep fertility*

Under the extensive management systems as practiced by nomads, rams are present in the flock year-round. Ewes are able to increase their liveweight and body condition during the green season and come into oestrus during the first part of the dry season; lambing thus takes place from November to February, so that subsequent lactation and growth of lambs coincides with the green pasture season. If, due to adverse rainfall conditions, feed requirements of the ewe are not met during this period, they do not come into oestrus or fail to conceive. They then lamb only once in two years. Under such conditions there is no advantage to high fertility in ewes. Consequently, these extensive breeds normally have single lamb litters. In such breeds, prolificacy can only be improved from 0.5 to 1 by improving nutrition during the green season, either by improving pasture availability or by supplementing the ewes before the beginning of the mating season (steaming up, see Section 4.4.5).

In improved sedentary situations inputs into the system are only warranted if prolificacy is improved by ewes lambing at least once every year and achieving a high twinning rate. In the temperate climates this is the rule and is easily achieved by using high fertility breeds and out-of-season breeding management practices; in the grazing systems of the arid zone it is the exception.

Table 4–1.
The nutritional value of feeds used at Migda for feeding sheep, expressed in Scandinavian feed units
(FU) per kg dm and in megajoules of metabolizable energy (MJ ME).

Feed	MJ ME per kg dm	FU	kg dm per FU	Source
Barley grain	12.9	1	1	ARC (1965)
Wheat grain	13.5	1.05	0.95	ARC (1965)
Mean green pasture	11.3	0.88	1.14	ARC (1965)
Mean early dry season pasture	9.59	0.74	1.35	de Ridder *et al.* (1986)
Dry season pasture	6.74	0.52	1.92	de Ridder *et al.* (1986)
Barley straw	6.5	0.50	2.00	Benjamin (1983)
Wheat straw	6.28	0.48	2.08	Benjamin (1983)
Poultry litter	8.37	0.65	1.54	Benjamin (1983)
Citrus pulp	11.97	0.93	1.08	Krol (1978)
Cotton waste	6.15	0.48	2.08	Krol (1978)

4.3.3 *Out-of-season breeding and lambing*

Out-of-season breeding is sometimes an attractive option because by January pasture availability in both quantity and quality can be predicted with reasonable accuracy. This also holds for agro-pastoral systems (Section 4.7.2), as in favourable rainfall years high pasture availability can be better utilized and in relatively dry years available forage quality may be high, due to the option of grazing drought-affected wheat with unharvested grain. Only in severe drought years will out-of-season breeding be disadvantageous because then heavy supplementation of ewes would be necessary.

Native breeds have a long and distinct anoestrus period from January to June. Consequently, there is little scope for increasing prolificacy by natural out-of-season breeding. It may be stimulated by hormone treatments, but even then only up to 60 percent of the ewes will conceive and those not conceiving may not recycle (Degen *et al.*, 1987). Crossbred sheep, however, exhibit oestrus up to March, allowing out-of-season breeding without hormones of ewes that failed to lamb up to then. Conception during this time results in lambing from June to September when only dry pastures of low quality are normally available. However, if pasture availability during the green season is high, even lactating crossbred ewes will conceive and have successful pregnancies. Most of the crossbred ewes that failed to lamb in the breeding season and up to 30 percent of ewes that have lambed, can lamb out-of-season (Degen *et al.*, 1987). In this way up to 30 percent of the ewes may lamb three times every two years, and this fraction can be increased by hormone treatment.

Lambs bred out-of-season may exhibit a lower growth rate because of a lower milk production and consequently a lower milk intake and must be weaned earlier than within-season lambs. This can be compensated for by lower mortality rates because of the warmer environment and earlier concentrate intake from creep feeding (see Section 4.5.6). These lambs can learn to drink water within 10 days (Barkai, 1979), because of the lower milk intake and higher ambient temperatures than in winter. Under this management, lambs generally are weaned at an age between 30 and 45 days at a liveweight of 12 to 15 kg and can achieve mean growth rates in the feedlot of up to 350 g per day, comparable to within-season lambs (Ungar, 1984).

4.4. Ewe nutrition

In this book the Scandinavian Feed Unit (FU) will be used to express the nutrient requirements of sheep. Even though this system has been replaced by more precise nutritional standards, it is used here because it is a single integrated feed unit that is sufficiently precise for a discussion that focusses more on system management aspects than on nutrition physiology. It is based on the feed value of barley grain for ruminants, 1 FU being equivalent to 1 kg of barley grain. The FU values of the feeds most commonly fed to sheep at Migda are given in Table 4–1.

4.4.1. *Maintenance requirements*

The maintenance requirement for a ewe of 55–60 kg in a feedlot is 0.6 FU per day (Krol, 1978). The requirement for grazing sheep can be up to 80 percent higher (Benjamin *et al.*, 1977), depending on pasture availability and quality. During the green season, pasture availability and quality are normally high, and maintenance requirements are relatively low, especially under deferred grazing. However, during the hot dry season sheep spend more time to select high quality plant parts whose availability gradually decreases (de Ridder *et al.*, 1986). Hence, maintenance requirements are higher than during the green season. After satisfying a hunger stress the animals prefer to rest under shade or to be near a drinking water source. Thus, under these conditions of low feed availability, intake is insufficient to meet maintenance requirements and the sheep gradually lose weight and body condition (de Ridder *et al.*, 1986).

The maintenance requirement measured for sheep grazing dry pastures in Migda was 0.64 MJ per kg metabolic weight (liveweight$^{0.75}$) (Benjamin *et al.*, 1977), equivalent to 0.86 FU for a 55 kg ewe.

4.4.2. *Flushing*

Body condition during in-season breeding can be improved by supplying additional feed to ewes that are below their mature liveweight and in poor body condition. Rapid increase in liveweight and body condition during the 45 days before mating promotes oestrus (Coop, 1966a). This practice is called *flushing*. However, Coop (1966b), has shown that flushing has no effect when ewes are

Table 4–2.
An index of body condition of ewes by subjective grading (Source: Jeffries, 1961).

Grade	Condition	Explanation
0	Emaciated	Details of bony structure easily seen
1	Poor store	Backbone prominent and sharp
2	Store	Backbone prominent but smooth
3	Forward store	Backbone smooth and rounded
4	Fat	Backbone and ribs can hardly be felt
5	Very fat	Backbone and ribs cannot be felt

at or near mature liveweight and in good body condition during the breeding season. Under intensified grazing management, ewes reach their mature liveweight and a good body condition at the end of the green season and normally stay in good condition during the early dry season and up to the beginning of the mating season. Hence, flushing is only necessary under unfavourable pasture conditions. Where flushing is required, 0.5 FU per day is given as a concentrate supplement during 45 days prior to mating. An index of the body conditions of ewes is given in Table 4–2.

4.4.3. *Breeding*

If body condition and liveweight are satisfactory at the onset of the breeding season, the level of intake during breeding is not critical. In-season breeding coincides with a time when pastures are dead and dry and mean quality is relatively low. However, because of selection for high quality parts, such as young leaves and seed heads, intake can initially meet maintenance requirements. Later, as availability of high quality material declines, intake gradually falls below maintenance requirements and the ewes gradually lose weight. Under such conditions, however, fertility is not adversely affected because the degree to which the ewes come into oestrus and conceive is dependent on nutrition in the preceding period (de Ridder *et al.*, 1986).

4.4.4. *Early pregnancy*

Nutritional requirements during the first 3.5 months of pregnancy are almost equal to maintenance requirements. Nutrition during this period is not critical, provided the ewe is in a satisfactory body condition. Pasture quality during this period generally decreases and though total dry matter availability may be high, intake may not meet maintenance requirements because of selective grazing. Ewes grazing dry pastures tend to spend most of their time selecting better quality herbage parts of low availability. Possibly, for this reason and the added stress of high ambient temperatures, sheep intake level may only satisfy a hunger drive rather than a satiation level. Sheep intake may thus be as low as

1.0 kg dry matter per day, and sheep will lose body condition and liveweight, but with no serious effect on the developing foetus or on the lamb at birth (de Ridder *et al.*, 1986).

4.4.5. *Late pregnancy and 'steaming up'*

The nutritional requirements of ewes increase considerably during the last six weeks of pregnancy (ARC, 1980; NRC, 1975) and the weight and viability of the lambs at birth and the initial height of the lactation curve are dependent on the level of nutrition during that period (Peart, 1968; Treacher, 1970). Depending on the degree of nutritional deficiency, the following disorders can occur: abortion, pregnancy toxaemia, weak lambs at birth, low milk production, death of lambs and/or ewes. Hence, this is a critical period coinciding with the end of the dry season and the beginning of the wet season. Supplementation may, therefore, be necessary, and when given at this physiological stage is called *steaming up*. Confining sheep to a holding paddock during grazing deferment is an advantage, as they can be fed there according to normative standards (ARC, 1980; NRC, 1975), using supplementary feeds of known nutritional value. As satiation intake is relatively low during this time and the requirements above maintenance can be up to 0.8 FU per day (ARC, 1980), the steaming-up ration is best given as concentrates with high digestible nutrient concentration.

4.4.6. *Lactation – first four weeks*

Lambing starts from late November and, depending on pasture availability, ewes may have to be supplemented to achieve intake above maintenance requirements. If ewes are confined during grazing deferment, they have to be supplemented according to normative feeding standards, *e.g.* ARC (1980).

The potential milk yield is determined by the genetic properties of the ewe and its nutrition. A typical lactation curve (from the data of Hadjipieris *et al.*, 1966) as given in Figure 4–2 shows that milk yield reaches a peak about 30 days after parturition; this holds also for mutton sheep in Israel (Folman *et al.*, 1966) and wool sheep in Australia (Corbett, 1968). It has been shown that peak milk flow is largely determined by the degree to which nutrition of the ewe meets its feed requirements for potential milk production (Barnicoat *et al.*, 1949a; 1949b). However, nutrition during late pregnancy can also influence the peak yield by affecting the timing of the onset of lactation following parturition, the ability to mobilize body reserves and the ability of the lamb(s) to draw out the milk (Treacher, 1970).

The general shape of the lactation curve cannot be influenced by nutrition after the first 30 days following parturition. Therefore, nutrition of the ewe during late pregnancy and early lactation is vital for growth and survival of the sucking lamb. As that period generally coincides with a time of low pasture availability (Fig. 4-1), grazing and supplementary feeding management has

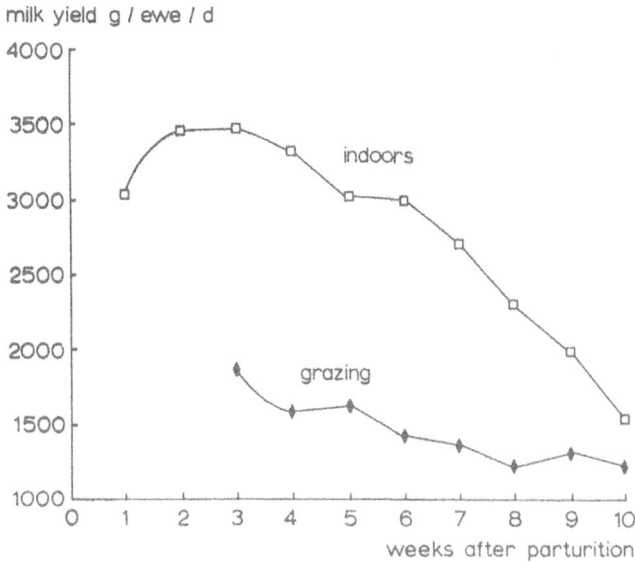

Fig. 4-2. The mean milk yield following parturition of ewes fed grass cubes indoors and of grazing ewes (Source: Hadjipieris *et al.*, 1966).

great importance in ensuring that nutritional requirements for achieving potential milk production are met.

The Finn cross sheep produce only enough milk to allow lambs to grow at a rate of approximately 0.25 kg per day, both under confinement (Barkai, 1979) and under grazing conditions (Benjamin *et al.*, 1980). These rates can also be achieved where creep feeding of lambs is practiced (see Section 4.5.6). Evidence from experiments with artificial rearers, using both milk substitute and ewe milk, has shown that lambs can convert the dry matter from milk with a dry matter content of 0.2 kg per kg into liveweight gain at an efficiency of 1:1 (Barkai, 1979; Large and Penning, 1967). Thus a mean liveweight gain of 0.25 kg per day corresponds to a mean milk intake of 1.25 kg per day. Measured values for lamb milk intake range from 0.9 to 1.3 kg per day (Benjamin *et al.*, 1980). The effect of low pasture availability on ewe dry matter intake and milk and pasture intake of their lambs is illustrated in Table 4-3. Lamb growth rates were highest when pasture biomass was greater than 650 kg dry matter per ha.

4.4.7. *Lactation – later stages*

The downward slope of the lactation curve and the length of the lactation period (Fig. 4–2) are determined by the level of nutrition during the period of increased milk production. However, if previous intake has met the requirements, this period is not critical, as pasture availability generally increases. Nutrition should be at least sufficient to maintain liveweight and

92

Table 4–3.
The intake from pasture by ewes and lambs at different biomass availabilities and the milk intake and growth rate of lambs (Source: Benjamin et al., 1980).

Biomass availability kg dm per ha	Ewe intake per day kg dm	Lamb intake per day		Lamb growth rate per day kg
		pasture kg dm	milk kg dm [1]	
790	2.20	0.41	0.29	0.29
690	2.02	0.46	0.29	0.27
610	1.96		0.27	0.24
580	1.50	0.63	0.16	0.23
490	1.49	0.55	0.14	0.20
480	1.80	0.43	0.22	0.23
450	1.71	0.37	0.22	0.24
420	1.60		0.18	0.16
390	1.57		0.17	0.21
310	2.02		0.26	0.20
300	1.98	0.36	0.27	0.25
250	1.95	0.23	0.31	0.25

[1] expressed per kg liveweight

body condition at such a level that fertility is not adversely affected. Lamb management provides flexibility during this period. Lambs that have reached 12 to 15 kg liveweight at 6 weeks of age (mean growth rate of up to 0.3 kg per day) can be safely weaned to a feedlot, if pasture availability is not sufficient to maintain high lactation levels or if pasture intake must be reduced to adjust to pasture growth rates. Weaning of the lambs will reduce the satiation intake of the ewes and a shorter grazing time will further reduce intake.

4.5. Lamb production

4.5.1. *Importance of lamb in the production system*

Lamb is the most important product sold from sheep production systems in many regions in the Middle East. The efficiency of conversion of quality feed falls with increasing liveweight and age (Graham and Searle, 1975). As concentrate feed given in feedlots is a major component of production costs, lamb nutrition is a critical factor in the grazing management of the flock.

4.5.2. *Pre-weaning nutrition up to sixty days of age*

There are important interactions between ewes and lambs in determining the lamb's intake level of mainly milk during the first six weeks of life and of milk and solid feed with increasing age and liveweight. The ewe determines potential milk availability, but the weight of the lamb and its capacity to suckle determine lamb intake. Small and weak lambs have difficulties sucking and without shepherd assistance may die. The stronger the lamb sucks, the more milk is produced, up to the ewe's genetic potential. In the period from birth to about 30 days of age, when the lamb is totally dependent on milk for survival and growth, a strong lamb can suck to appetite and achieve a growth rate of more than 0.3 kg per day. At six weeks of age, lambs can be weaned and survive on solid feeds (Benjamin *et al.*, 1981).

4.5.3. *Artificial rearing*

Sheep breeds with a high prolificacy tend to have a lower genetic potential for milk production and often have difficulty raising all the lambs that are born. In a flock with a prolificacy of over 1.5, ewes may not be able to raise up to 20 percent of the lambs born. In flocks where out-of-season breeding of hoggets is practiced, this proportion may even be higher. Such lambs can be raised in an artificial lamb rearing system.

Artificial rearing systems are based on the use of a commercial milk substitute and a system for delivering the milk to the lambs. The milk substitute commonly has a fat content of 20 to 30 percent and is cheaper per unit dry matter than either ewe or cow milk, but is more expensive than concentrates. The milk is hand-mixed to the correct consistency, using either cold or warm water. The delivery system in its simplest form can be a bucket with self-sealing teats at its base. More sophisticated artificial rearing systems are available that automatically mix the milk substitute with water and maintain the resulting product at a constant temperature. They can accomodate up to 200 lambs and are available at a cost of around $ 3,000. The simple low-cost systems are usually used on farms lacking electricity and the more expensive systems where large numbers of lambs must be reared artificially and where labour costs are high.

Artificial rearing entails additional labour costs: milk must be prepared, a high level of hygiene kept, and all lambs must be taught to suck from the teats. The experience in Israel is that one person can handle 200 lambs with sophisticated systems.

4.5.4. *Pre-weaning nutrition after sixty days of age*

From sixty days of age onwards, lamb milk intake decreases almost linearly and intake of solid feed increases similarly (Hodge, 1966). Feed from high quality green pastures has a crude protein content of about 18 percent, similar to that of milk, and if pasture availability is high enough, lamb growth rates can be maintained at 0.2 to 0.3 kg per day or even more. If pasture quality is lower and/or availability becomes limiting, growth rates will decline. In such a situation it usually becomes necessary to wean the lamb.

4.5.5. *Post-weaning nutrition from pasture*

The objective of good weaning management is a transition from a mixed diet of milk and solid feed to a diet of solid feed alone, with a minimum check in the growth rate of the lambs. This can be achieved if, prior to weaning, intake of solid feed from high quality pasture already comprises a significant proportion of the diet and if after weaning high quality pasture is still available. Ideally, lambs should be weaned onto pasture with a crude protein content of 12 to 18 percent. Under such conditions, lambs weaned at 12 to 15 kg liveweight can achieve growth rates of more than 0.2 kg per day (Benjamin *et al.*, 1981). If, after weaning, the ewe is in good body condition and her nutritional requirements are at maintenance level, weaned lambs should have priority for grazing in the better quality pastures. Dry ewes can follow, after the lambs have selectively grazed leaves.

Special purpose high quality weaning pastures can facilitate successful weaning. Annual legume pasture (*e.g. Medicago polymorpha* L.) can be heavily stocked with lambs during this period so that only a relatively small area may be needed (Benjamin *et al.*, 1981). However, this pasture is not without a cost that must also be considered in relation to other feeding options (see Chapter 6). Because of the variability in the length of the pasture growing season, lambs may not always be able to achieve target sale weights in the pasture, in which case provision must be made for concentrate feeding. This is best achieved if both before and after weaning on to pastures, creep feeding has been practiced (see Section 4.5.6).

4.5.6. *Lamb fattening from concentrates in feedlots*

Intensive lamb fattening is possible in semi-arid mediterranean regions because lambs weaned onto concentrate rations can achieve a mean ratio of feed intake to liveweight gain of 5 to 1 during a liveweight increase from 15 kg to 45 kg (Folman *et al.*, 1976). This is economically viable whenever the price ratio of lamb liveweight to concentrates is higher than 10 (Benjamin and Harel, 1983).

The length of the green pasture season in any particular year determines the age and liveweight at weaning. Weaning weights usually vary between 15 kg and 30 kg. The market preference is for lambs weighing 35 to 45 kg. This can be attained by feeding concentrates to weaned lambs under feedlot conditions. Lambs cannot be weaned directly from pasture to a concentrate ration, because of the danger of overeating and acidosis, which may lead to death. Even when lambs are gradually accustomed to concentrates over a ten day period, there is an undesirable growth check.

To avoid these problems *creep feeding* of lambs before weaning can be practiced. Under this system concentrates are made available to the lamb *ad libitum* from birth, using a partitioning that allows access to the concentrates by the lambs but not by the ewes. In that way lambs can achieve a concentrate intake of over 0.2 kg per day before being weaned and transferred to an all concentrate fattening ration.

The use of creep feeding can delay weaning as the concentrate intake compensates for lower milk and pasture intake, so that relatively high growth rates can be maintained for a longer period of time. Moreover, conversion of concentrates to lamb liveweight is efficient, because the basic requirements for maintenance and some liveweight gain is still being met by milk and pasture (Barkai *et al.*, 1981).

Creep feeding allows another fattening option. If pasture availability and quality are high, lambs can be weaned onto pasture while continuing *ad libitum* concentrate feeding. In that case feed requirements for maintenance can be met partially or completely from pasture intake, resulting in a more efficient conversion of concentrates to liveweight gain. Liveweight gain may, however, be lower than under feedlot conditions.

4.6. Flock management

4.6.1. *Culling and replacements*

In a lamb production system infertile sheep are non-productive and decrease flock performance. In intensive systems, where expensive inputs are used, such ewes reduce profitability and should be culled from the flock and sold. Ewes must also be culled for age. Normally, even in extensively managed flocks, ewes are culled at six years of age, regardless of the total number of lambings. In intensively managed systems, ewes may have lambed 7 times at 5 years of age and are likely to be weaker than ewes of the same age in extensive systems. The probability of damaged udders, lower liveweight and poorer body condition is also higher, all of which can result in lower fertility and higher mortality.

Taking into account 5 percent mortality per annum, a normal value under grazing conditions in Israel, the replacement rate for an intensively managed flock is approximately 25 percent and for an extensively managed flock approximately 20 percent. This means that at all times the flock is composed of 20 to 25 percent ewe hoggets and 75 to 80 percent mature breeding ewes.

4.6.2. *Ewe hogget breeding*

After weaning, ewe lambs (hoggets) from the prolific breeds can reach sexual maturity at an age of 6 to 8 months if their liveweight is over 40 kg by that time (Goot *et al.*, 1976). Ewe hoggets may comprise up to 25 percent of the flock, depending on replacement rate, so that normally in any one year only 75 percent of the flock can be fertile. Hence, the ability to breed ewe hoggets in their first year is an important aspect of improved sheep production.

At Migda all lambs are usually fattened under feedlot conditions so that ewe lambs can achieve 40–50 kg of liveweight at the age of 6–10 months and can then be mated with ram lambs from the same feedlot, which also mature sexually at the same time or even earlier (Goot *et al.*, 1976). During 4 years of study it was shown that at least 60 percent of the fattened lambs conceived and could be identified by ultrasound pregnancy testing. Pregnant hoggets can thus be selected for replacements in the flock and surplus hoggets can be sold for mutton.

The nutritional management from weaning to 50 kg liveweight is identical to that for fattening and from 50 kg liveweight onwards it is similar to that for mature ewes kept in confinement during autumn and early winter under deferred grazing. Ewe hoggets tend to lamb somewhat later and have a slightly lower fertility than mature ewes. This later lambing may be advantageous as the hoggets are then more likely to begin lactation when pasture availability is high. Intake can then reach satiation levels which exceed requirements for maintenance and lactation, enabling the animals to increase body weight at least up to their potential mature liveweight so that at the beginning of the next breeding season, they can conceive successfully and achieve the potential fertility of the breed.

4.7. Pasture productivity and systems of utilization

4.7.1. *System intensification*

In unimproved natural pastures with relatively low production, native breeds of sheep with low prolificacy, like the Awassi, predominate. Sheep performance can be improved only to a limited extent up to the potential prolificacy of the breed by using relatively cheap inputs to improve animal health and reduce mortality from contagious diseases. Where grazing is limited to natural pastures, such animal husbandry systems are defined as *extensive*.

Higher production from improved pastures can only be fully utilized by intensified sheep production systems. At a low stocking rate feed availability per animal may be above optimum and feed can be wasted. Increasing the stocking rate has important consequences for (i) the quantity and quality of the herbage produced, (ii) the proportion of herbage produced that is consumed, and (iii) the herbage intake of individual sheep (Willoughby, 1959; Morley and Spedding, 1968). These interactions are discussed elsewhere (see Chapters 2 and 6). In general, sheep production per unit area continues to increase with increasing stocking rate beyond the point where production per head begins to decrease (Mott, 1960; Jones and Sandland, 1974). This is because even though the intake per sheep decreases, the production from the increased number of sheep more than compensates for the resulting lower production per sheep. However, in semi-arid regions fertilizer application, while increasing average pasture production, also increases interannual variation in pasture availability (van Keulen, 1975; van Keulen *et al.*, 1981). Hence, there is a danger of lower production under high stocking rates if they are not supported by supplementary feeding. Systems with higher stocking rates and increased inputs of fertilizer and feed supplements are defined as *intensive*. Further increasing stocking rate can lead to lower lamb birth and weaning weights (Spedding, 1970). This creates greater reliance on supplementary feeds, such as milk substitutes to supplement weak lambs at birth and concentrates to fatten weaners to a marketable liveweight. This intensification requires costly inputs at all levels of the production system and, in addition, management expertise to support the demands of such a highly sophisticated production system. Increased stocking rates, higher ewe fertility, increased use of supplementary feeds and more labour per ewe, characterize *highly intensive* systems.

In Israel, these highly intensive systems are confined to the feedlot. Such systems use highly prolific breeds, practice early weaning and use hormones to induce oestrus 45 days after lambing. Feedlot rations can be fed in accordance with sheep nutrition requirements for the various production functions (ARC, 1965; ARC, 1980; NRC, 1975; NRC, 1985) in contrast to the grazing situation, where sheep intake is not regulated by their requirements but by the level of pasture availability and its quality (Noy-Meir, 1978). These highly intensive systems can be profitable as long as the price ratio between lamb liveweight and concentrate feed on a per kg basis is higher than 15.

4.7.2. *Agro-pastoral systems*

Favourable price ratios do not always prevail, so that often the intensive feedlot systems of sheep production are not profitable. With falling wheat prices, wheat cultivation in marginal areas like the northern Negev is also difficult to sustain economically. Wheat yields are marginally better in a wheat-fallow rotation than when wheat is grown continuously, but then half the area is non-productive every year. In an integrated sheep-wheat system, wheat production could possibly be justified, if pastures replaced the fallow, and wheat crops were grown in rotation with the pastures.

The integration of pasture and grain crops allows a more intensive pasture utilization. In favourable rainfall years aftermath stubble grazing provides adequate feed during summer from the end of May to the end of November. Hence, stocking rates high enough to fully utilize the pasture herbage by the end of May are feasible. In poor rainfall years, pasture production will not be sufficient to sustain such high stocking rates. However, the area available for grazing can be increased by adding the grain crop area. Dry matter production of the pasture and grain crops will be similar (van Keulen, 1975) and their combined yield may be equal to the dry matter available for grazing in the pasture area in an average rainfall year. By using such a system, the risk of overstocking in dry years is substantially reduced, although possibly at the expense of a grain yield.

The possibility of grazing the grain crop early in the season allows deferment of grazing in the natural pasture, which may be necessary at high stocking rates due to the pasture's slow initial growth rate. This grazing may even have a

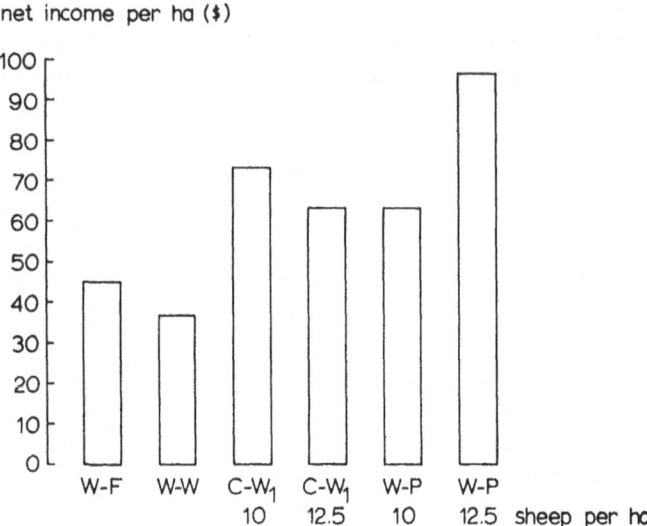

net income per ha ($)

Fig. 4-3. A comparison of the net income ($) of combinations of wheat and agro-pastoral systems over the years 1976/77 to 1979/80 inclusive, all fertilized with 100 kg ha^{-1} N and 26 kg ha^{-1} P; W-W = continuous wheat; for further abbreviations see Table 4–4.

positive effect on grain yield as it promotes early tillering, allowing the development of secondary tillers for grain (Seligman *et al.*, 1981) . This early grazing of grain crops can partially compensate for deferment of pasture grazing that is required when stocking rates are high (see Chapters 2 and 6).

Where only natural pastures are grazed, enough herbage has to be left during the growing season to meet sheep requirements during the dry season from May to November. In addition, even if aftermath grazing is available during the dry season, grazing of natural pastures during the growing season must be limited to allow re-seeding of the pasture. In both cases, up to half the green pasture of initial high quality must remain as stem-cured hay at a quality level not much higher than grain aftermath, which at least in quantity can meet sheep requirements during the dry season.

Where pastures have to be re-seeded under semi-arid mediterranean conditions and used in the same season, sometimes for one year only, small grain species like barley are most successful. It is not a pasture species in the strict sense of the word (Hodgson, 1979), but is relatively drought resistant, tillers early, has good winter production, grows relatively well under grazing and, if lightly grazed during the green season, produces some grain that improves the quality of the forage early in the dry season.

Two experimental agro-pastoral systems were set up at the Migda Experimental Farm to test the performance of such integrated systems. In one, a pasture of native, spontaneously occurring annual species was maintained for 5 years on half the area, and continuous wheat sown on the other half. This area was divided into five sections, each of 9.6 ha, and grazed with Merino-Awassi-

Table 4–4.
Rainfall and net income [1] per ha of system at Migda in fields of wheat and agro-pastoral systems,
fertilized with 100 kg N and 26 kg P per ha. Data from 1976/77 to 1982/83.

Year	Rainfall (mm)	Net income ($ per ha)			
		Wheat systems		Agro-pastoral systems	
		W–F	C–W	C–W1	W–P
1976/77	212 B	2	5	–50	20
1977/78	159 B	–64	–142	–82	–82
1978/79	200 B	–14	–120	–79	20
1979/80	368 G	256	402	350	219
1980/81	302 G	142	201	252	451
1981/82	203 G	122	195	97	210
1982/83	373 G	327	449	470	245
Mean 7 years	260	110	141	159	155

Abbreviations:
B   Unfavourable rainfall distribution
G   Favourable rainfall distribution
W–F   Wheat-fallow rotation in a wheat system
C–W   Continuous wheat in a wheat system
C–W1   Continuous wheat in a 'fixed ley' agro-pastoral system;
           50 % of the area in natural pasture with a stocking rate of
           5 ewes per ha of system
W–P   Wheat/sown pasture in an 'alternative ley' agro-pastoral
           system; 50 % of the area in sown pasture, with a stocking
           rate of 7.5 ewes per ha of system
[1]   All cultivation costs and income prices adjusted to those
           of 1976/77 at Migda.

Finn cross-bred sheep of medium prolificacy (1.3 lambs per ewe) at stocking rates of 3.3, 6.6, 10, 12.5 and 15 sheep per ha of pasture, respectively. This system was designated 'semi-intensive'. In a second system, a barley pasture and a wheat crop were sown annually in a two year rotation. This was designated 'intensive' and as in the previous system, was subdivided into five 'farmlets'. Each farmlet was stocked with 20 Merino-Finn sheep of high prolificacy (1.8 lambs per ewe) and the area was adjusted to create stocking rates of 10, 12.5, 15, 17.5 an 20 sheep per ha of sown pasture, respectively.

In both systems, pasture and wheat were fertilized annually with 100 kg N per ha as urea and 26 kg P as granulated superphosphate. In the semi-intensive system the pasture was integrated with the wheat in such a way that grazing of wheat aftermath was the rule and the green wheat was grazed only occasionally for a short period before ear differentiation. Every effort was made to harvest a grain and straw crop. Only in very unfavourable rainfall years the green wheat fields were grazed completely. In the intensive system wheat was more fully integrated with the pasture, as the wheat was almost always grazed before ear differentiation and could be grazed at other times, if necessary, to meet sheep feed requirements.

In both systems ewes lambed once a year, but the supplementary feed regime varied. In the intensive system sheep were given a flushing ration of 0.5 FU per day (see Section 4.4.2) for 45 days, but not in the semi-intensive system. The rations during steaming up (see Section 4.4.5) and grazing deferment were 0.5 FU per head per day higher in the intensive system.

In Table 4–4 and Figure 4–3 the financial results of both systems are compared with those of the wheat-fallow and the continuous wheat systems described before. In Table 4–4 the most profitable stocking rate in each agro-pastoral system (5 and 7.5 sheep per ha in the semi-intensive and intensive system, respectively) is compared with the two wheat systems over a 7 year period which included 3 unfavourable rainfall years. The results show large fluctuations from year to year in all systems, as expected, but over the experimental period as a whole, the agro-pastoral systems were more profitable than the two wheat systems. In Figure 4–3 two stocking rates in the agro-pastoral systems (10 and 12.5 sheep per ha of pasture are equivalent to 5 and 6.25 sheep per ha of system) are compared with the two wheat systems over the first 4 years of the experiment. These results also show the financially better performance of the agro-pastoral systems and, in addition, that in the semi-intensive system the lower stocking rate gave better results, while in the intensive system the opposite occurred.

5. Description of a pasture system generator (PSG)

N.G. SELIGMAN

5.1. General outline

Sheep, pasture and dryland wheat farming can be combined in many different ways and at different levels of intensification. The combination that is possible or prevalent in any region is determined by many factors, among which physical and socio-economic conditions predominate (Bowden, 1979; Carter, 1981; Donald, 1965; Noy-Meir and Seligman, 1979). As conditions change, and especially when active development is undertaken, there can be far reaching changes in the type of operation. Identification of feasible development pathways is necessary for rational planning of regional agricultural development. The region is composed of various farming systems that can be implemented in various combinations within and between separate farms. The purpose of the pasture system generator (PSG) described here is to define a large number of such production systems and so provide a sufficiently wide range of technological options facilitate relatively unrestricted analysis of very different development scenarios (de Wit *et al.*, 1988). These can vary so as to favour very extensive, very intensive or intermediate levels of development. Examples of all levels of agro-pastoral intensification can be found in the semi-arid regions of the world today and even in the more restricted region around the Mediterranean Basin.

The systems defined by the PSG represent a series of technically feasible input/output relations that vary with regard to the level of input intensity and with regard to the genetic stock (= sheep breed) appropriate to that intensity. An individual farmer can implement one such system, or in some cases a combination of two or more. Thus, for instance, the flock may be composed of two breeds with different performance levels and the ratio between them can change with time in accordance with the breeding goals. Within a region the range of systems implemented is generally larger because of the greater diversity of operators and operating conditions. The PSG defines pastoral systems. Consequently it deals mainly with the management and productivity of the pasture and with the maintenance, feeding and breeding of the flocks, including flocks raised in feedlots. Grain cropping systems are defined separately, but are eventually analysed together with the pastoral systems (see Subchapter 5.3).

Most producers within the mediterranean region have access to supple-

Th. Alberda et al. (eds.), Food from Dry Lands, 101–132.

mentary feed and so the systems are defined on the assumption that animal feed requirements are met from pasture, wheat aftermath grazing, straw, and/or concentrate feed according to the intensification level (Noy-Meir, 1975b; Noy-Meir and Seligman, 1979). Where subsistence systems are defined, production targets are set to lower long-term levels that can be maintained on pasture resources and emergency feeding only. Thus, all systems are characterized by an attainable production level. The input/output relations are calculated by first defining a series of production levels and the resource requirements necessary to attain those levels. These levels can be attained by different input combinations, management competence and breed selection according to ewe fertility (Spedding, 1970). The PSG calculates the specific inputs that are necessary to attain a given production level for a set of appropriately defined production systems. The procedure is target-oriented in the sense that the production level (rather than the input level) is predetermined and the inputs necessary to obtain the production target are calculated accordingly (Spedding, 1975).

Capital spending and investments increase with greater intensification as more associated inputs like sheds, fences and veterinary costs are needed. When flock nutrition becomes stabilized with feed stored or bought from outside the farm, a more prolific breed of sheep may be justified. Fertilizers can produce more feed and allow a larger flock and more intensive land utilization. In the most intensive livestock operation, the feedlot, all the feed is given as concentrate or roughage. The decrease in dependence on the herbage production of the pasture is accompanied by a greater dependence on resources outside the system or the region.

The time unit of the PSG is one year and all calculations eventually lead to an annual balance. The pasture productivity data refer to the semi-arid mediterranean area and are average data taken over a 5 to 10 year period in Israel.

5.1.1. *Range of system variation*

A large range of systems is defined so as to avoid overlooking desirable combinations that have relevance to reality. In general terms the systems can be described as follows:
- *Extensive systems*, where the basic breed is the local Awassi fat-tail sheep, including improved selections; flock nutrition is mainly based on the available feed resources (pasture and wheat crop residues); lambing rates are no higher than 100 percent, generally less, but seldom lower than 50 percent (Epstein, 1985; Noy-Meir and Seligman, 1979).
- *Intensive systems*, where more fertile breeds are used, mainly the German Mutton Merino and various crosses; nutrition depends more heavily on concentrate supplement; net lambing rates are between 100 and 160 percent.
- *Highly intensive systems*, where very fertile breeds are used, mainly Finn and/or Romanov crosses; nutrition is heavily dependent on supplementary

concentrate feed; net lambing rates are between 160 and 240 percent, thus including not only multiple births, but also three pregnancy cycles in two years. The most intensive systems have also early weaning and, in some cases, rearing of lambs in labour saving artificial rearers.

These systems span the following range of production possibilities:

– Level of production based on breeds with prolificacy from 0.5 to 2.4 weaned lambs per ewe per year, with birth weights from 3.0 to 5.5 kg, weaning weights from 15 to 30 kg and sale weights from 20 to 50 kg per lamb.

This range of sheep production systems can be expanded to include a range of management and land use systems based on cultivable land that can be used for grain cropping or pasture, or on rangeland that can be used for grazing only.

The management variables that are defined include:

– Supplementary feed based on concentrate feed or on roughage.
– Pasture: fenced or not; fertilized or not.

Pasture for very extensive systems is regarded as not suitable for chemical fertilization, whereas intensive systems could be based on fertilized pasture to the extent that they are based on pasture use at all. Pasture can be grazed during the green season only or yearlong.

– In intensive systems an artificial weaner can be included or not.
– Structures like corrals or lambing pens can be built or not; they are generally more complex and expensive the more intensive the system.

5.1.2. *Technical characteristics of the system*

The full formal complement of combinations between sheep production, feeding and land management systems would include unlikely or unfeasible combinations, like highly intensive sheep production systems based on extensive rangeland. The set that was chosen as technically realistic was subdivided according to the following criteria:

– Cultivable land, which allows a wide range of possibilities, including the whole range of sheep production systems, that could be fenced or not, with a feeding system based on either concentrate or roughage.
– Rangeland, which allows a narrower range of management options, including sheep production systems which are not highly intensive. It is also assumed that if management possibilities are low and fencing is not available, concentrate feeding would not be likely. Consequently, rangeland systems can be fenced or not, but when not fenced, then the heavier concentrate feeding option is also not defined.
– Finally, sheep production systems that are based on complete maintenance in the feedlot, are unlikely to be extensive. Consequently, the feedlot option includes only intensive and highly intensive sheep production systems.

A full description of the characteristics of the chosen set of systems is given in Table 5–1. The following descriptions are not all mutually exclusive and some, which are redundant as discriminators between systems, are included to fill out

Table 5-1.
System definition by pasture characteristics and feed supplementation method.

group number	land type	fencing	main feed for supplementation	system identification
I	cultivable	unfenced	roughage	001 - 018
II	cultivable	unfenced	concentrates	101 - 118
III	cultivable	fenced	roughage	201 - 218
IV	cultivable	fenced	concentrates	301 - 318
V	range	unfenced	roughage	401 - 411
VI	range	fenced	concentrates	501 - 511
VII	no pasture	-	roughage	412 - 418
VIII	no pasture	-	concentrates	512 - 518

the definition of the system. Thus, for instance, birth weight does not discriminate between systems, but describes some in greater detail. These details are later used in calculating the input/output relations.

The descriptive characteristics are:
- the breed of sheep, which can be either the Awassi (Awas), the Improved Awassi (Impr), the German Meat Merino (GMM), or the Finn Cross (FinnX);
- the grazing system, which can be yearlong grazing (yearl), yearlong except for deferment during a transitional period after the start of the winter rains (defer), or green season grazing, including the early dry season with high quality pasture (green), or feedlot systems with no grazing at all (-);
- the weaning system, with normal weaning (norm), early weaning (early), or early weaning plus an artificial rearer (artre);
- the system intensity: extensive systems (ext), intensive systems (int) and highly intensive systems (hint);
- the net lambing rate, ranging from 0.5 to 2.4 weaned lambs per ewe per year;
- the birth weight, weaning weight and sale weight of lambs in kg per lamb;
- the fertilizer application option (0 = no; 1 = yes).

Table 5-2 gives 18 technically feasible sheep production combinations. In combination with Table 5-1 there are in total 108 different systems which cover most of the practical possibilities in Israel today. For other regions the number of systems, the marketable products, as well as the ranges applied may be different.

5.1.3. *Some general comments*

The inputs in each system include pastures, supplementary feed (concentrates and roughage), fertilizer, veterinary costs, miscellaneous current costs (including fuel, various materials, levies, etc.), and capital investments in fences, stock watering points, corrals, lambing sheds and equipment.

Table 5–2.
Pasture system definition by level of lamb production (or ewe prolifacy), grazing method and sheep characteristics.

				System parameters [1]						
A	B	C	D	E	F	G	H	I	J	K
1	Awas	yearl	ext	norm	lev.1	0.50	3.0	20	20	0
2	Awas	green	int	norm	lev.1	0.50	3.0	20	20	0
3	Awas	yearl	ext	norm	lev.1	0.65	3.5	25	25	0
4	Impr	yearl	ext	norm	lev.1	0.75	3.7	33	33	0
5	Impr	green	int	norm	lev.1	0.75	3.7	33	33	0
6	Impr	yearl	ext	norm	lev.2	0.90	4.0	40	40	1
7	Impr	green	int	norm	lev.2	1.00	4.5	40	45	1
8	GMM	defer	int	norm	lev.2	1.00	4.5	40	45	1
9	GMM	defer	int	norm	lev.2	1.20	4.5	40	50	1
10	GMM	defer	int	norm	lev.3	1.40	4.0	40	50	1
11	GMM	defer	int	early	lev.3	1.60	4.0	30	50	1
12	GMM	green	hint	norm	lev.3	1.70	3.8	30	50	1
13	GMM	green	hint	early	lev.4	1.90	3.8	20	50	1
14	GMM	green	hint	artre	lev.5	1.90	3.8	15	50	1
15	FinnX	green	hint	norm	lev.4	2.00	3.5	20	50	1
16	FinnX	green	hint	early	lev.4	2.20	3.5	15	50	1
17	FinnX	green	hint	early	lev.4	2.40	3.2	15	50	1
18	FinnX	green	hint	artre	lev.5	2.40	3.2	15	50	1

[1] Definition of system parameters:

A   System identification number.
B   Sheep breed: Awas = Awassi; Impr = Awassi under improved management; GMM = German Mutton Merino; FinnX = Finn cross.
C   Grazing system: yearl = yearlong; green = green season only (includes early dry season when quality of pasture is still high); defer = yearlong except for deferment during transitional period after opening winter rains.
D   System intensity: ext = extensive; int = intensive; hint = highly intensive.
E   Weaning system: norm = normal weaning; early = early weaning; artre = early weaning plus labour saving artificial rearer.
F   Technical know-how required: lev.1 - lev.5, low to high level.
G   Net lambing rate: lambs weaned per ewe per year.
H   Lamb birth weight: kg lamb^{-1}.
I   Lamb weaning weight: kg lamb^{-1}.
J   Lamb sale weight: kg lamb^{-1}.
K   Fertilizer application to pasture: 0 = no; 1 = yes.

The area of pasture allocated to each ewe depends on the pasture productivity and on the fraction of pasture in the flock feed balance. This fraction is a function of the length of the green and dry seasons in any particular system, as well as of the flock feed requirements when at pasture. Pasture productivity is an exogenous variable that is a mean value, related to the mean rainfall in the zone, to the land type and to the level of fertilization. This value can either be established on the basis of experimental values or may be calculated with the type of simulation models described in Chapter 3.

The total amount of feed needed to obtain the production target must be provided either by pasture (green and dry) or by supplementary feed (concentrates and roughage). The lower the production target, the easier it is to fulfil the requirements from pasture only, as the productive period of sheep can coincide with the productive pasture phase. As sheep management intensity increases, feed requirements exceed pasture resources and increasing reliance on supplementary feed is necessary.

Sheep nutrition requirements for the physiological functions involved in lamb production are derived from research and extension work in the region as well as from the animal nutrition literature. As long as the ewes are on pasture, it is assumed that they can obtain their maintenance needs on an annual basis, even though liveweight may fluctuate between poor and good pasture seasons. Mature ewes may lose weight during the dry months, but that is made up during the green season, so that from year to year mean ewe weight in the flock does not change very much. Feed requirements for lactation, lamb fattening, breeding, late pregnancy, as well as requirements for raising hoggets and maintenance of rams, are calculated. Whatever can not be supplied by the green pasture must be supplied as supplementary feed. The supplementary feed requirement is then allocated between concentrate and roughage, depending on the type of feed needed for the various physiological functions and on the relative availability (or cost) of concentrate and roughage.

Where the nutrition of the flock is based mainly on pasture and on wheat stubbles, the flock is subjected to periodic nutritional deficiencies because of the cyclic nature of feed availability and quality from these resources. In extensive systems there may be occasional supplementation, but in the more intensive systems feed deficiencies will be made up by supplementary feed as a matter of routine, mainly during the breeding (flushing) and lambing (steaming up) seasons. In the highly intensive systems all feed deficiencies for breeding and liveweight gain are made up by supplementary feeding.

The output of the flock, marketable lambs, is a function of ewe prolificacy and the target sale weight of the lamb. The actual number available for sale depends on whether the flock size is increasing or not, and, consequently, on whether the female lambs are grown to hoggets or sold for meat. This is determined exogenously in the PSG but endogenously in the 'technology selector' (see Chapter 7).The prolificacy, or net lambing rate, which is related to the breed and management intensity, also determines the selection level for the breeding stock.

System intensity determines the animal husbandry methods. In extensive systems the lambs are dropped in late autumn, early winter and are weaned soon after the end of the green season. Lambs are sold without feedlot fattening. The lambs that are not kept for replacement are generally sold after weaning. In intensive systems, lambs are weaned earlier and fattened to market weight mainly on concentrate feed. In highly intensive systems, hormones are often used to initiate out-of-season oestrus, and in this way increase the lambing frequency from two to an average of three lambings per ewe in two years (Goot *et al.*, 1976).

The labour requirements depend on system intensity and to a large degree on whether the pasture is fenced or not. The type of supplementary feed can be varied to a certain extent, depending on the availability of agricultural wastes, including straw and wheat stubbles. Fertilizer application to pasture to lengthen the grazing season and to increase pasture production is considered only for intensive systems (Benjamin *et al.*, 1982).

Investments in sheds and equipment are very low in the extensive systems, but increase sharply as the system intensity increases and as the net lambing rate increases. Investments in stock drinking water depend mainly on the size of the flock. In this study the size of the flock in turn is determined by the number of ewes that one person can manage at a given level of production intensity.

5.2. Calculation of system inputs and outputs

A set of functional relationships will be defined that constitute the routine by which the PSG, as described in the outline above, is implemented. The equations that are derived from these definitions are given in Table 5–3 and the Appendix of this chapter.

5.2.1. *Pasture*

In mediterranean-type pastures with a strongly seasonal growth pattern, the year is divided into a relatively short green season during which high quality pasture is available, and a long dry season during which the feed quality of the pasture vegetation is low and decreases with time. The length of the growing season and the amount of biomass produced are, therefore, attributes that can define the main determinants of pasture productivity for a given vegetation type (Noy-Meir and Seligman, 1979). Annual fluctuations can be wide and tend to increase as the amount of annual rainfall decreases (Le Houérou *et al.*, 1988; Seligman and van Keulen, 1989). In a target-oriented pastoral system the production goal is set to a level that is reasonably attainable in most years, as indicated by available experience. The 'reasonably attainable' level for a given region will be determined not only by the productivity and reliability of the pasture, but also by the availability and economic feasibility of supplementary feeding. The productivity of the pasture, in turn, can be influenced by the intensity and time of grazing (Willoughby, 1959). In particular, heavy grazing early in the growing season can severely reduce pasture productivity (see Chapter 2). In order to prevent loss of pasture production under such conditions, it may be necessary to defer the use of pasture until it can be safely grazed without fear of serious loss (Noy-Meir, 1975a, 1978; see also Chapter 2).

In the following description all systems are target oriented in that the expected production level for a region is the point of departure in defining the system (Spedding, 1975). Where supplementary feed is available, expectations can be set to a higher level, especially where superior genetic stock is present. In intensive systems most of the pasture can be utilized even in good years, so that a long-term average productivity can be a realistic basis for system definition. In subsistence systems, where supplementation is minimal, expected production levels will be determined by the production in poor years rather than by average production. As these conditions vary from region to region, they are defined in the PSG as regional characteristics. The feed balance of the 18 pastoral systems is consequently based on the long-term pasture productivity for the different land types in a particular region.

The annual pasture cycle can be divided into four phases:

a. The early green season, between the first germinating rains and the establishment of a pasture capable of continuing net growth at the stocking density characteristic for a given system. This phase is particularly important in intensive and highly intensive systems, because for these systems the grazing season does not begin before the end of this phase. In extensive systems, where grazing is yearlong, the end of this phase marks the beginning of the 'effective green season' when the animals can obtain all the necessary feed requirements from the production of the pasture. Generally, the lower and more erratic the rainfall, the longer will be the period of early season uncertainty. For the northern Negev, the period between the initial rains and the beginning of the effective green season is on the average around 80 days. For drier regions it will be longer, with about 100 days being the limit in the drier part of the Negev (Noy-Meir, 1975b). Fertilizer application can shorten the early season phase by one to four weeks (van Keulen, 1975).

b. The effective green season is the period during which pasture availability and quality are sufficient to meet all the nutritional requirements of the flock. This definition assumes that the stocking rate and grazing management do not reduce the expected pasture production. In order to ensure that this assumption holds, stocking rate and grazing management (*i.e.* the length of the early season grazing deferment) are derived from the production level and the length of the green season appropriate to the region under consideration. For a given region the length of the effective green season is derived from available data (*e.g.* Noy-Meir, 1975b; Zaban, 1981). It is not only dependent on rainfall conditions, but also on local differences in land type; cultivable land with a higher water holding capacity has a longer effective green season than the surrounding hill range land with shallow soils. As a rule, the length of the effective green season on the hill range is taken as half that of the cultivable area (Noy-Meir, 1975b). Fertilizer application on pasture is assumed to increase the length of the effective green season by one to four weeks (Benjamin *et al.*, 1982; van Keulen, 1975).

c. The early dry season, during which the pasture quality is lower than in the green season, but considerably higher than during the rest of the dry season. The period is defined as half the length of the green season, although it may also depend on utilization intensity (de Ridder *et al.*, 1986).

d. The main dry season, which is the period between the end of the early dry season and the first rains. During this phase, the value of the residual dry feed in the pasture is usually reduced to a very low nutritional value. The length of this period depends on system intensity. For extensive systems it is the period between the end of the early dry season and the effective green season; for intensive systems the period of deferment has to be subtracted; in highly intensive systems there is no grazing in the main dry season and grazing is limited to the duration of the early dry season.

Pasture utilization is dependent on system intensification level. Yearlong utilization is common in relatively extensive systems, where the fluctuations in pasture quality from season to season are reflected in a relatively low target sale weight (Noy-Meir, 1975b). In highly intensive systems use of low value poor quality dry pasture may not cover the cost of labour for maintenance of fencing or for herding the flock. There are also intermediate systems, where the dry pasture is used fully, but where deferment early in the green season is necessary to ensure both adequate animal nutrition and pasture development. The latter practice is particularly necessary if the stocking rate increases and when excessive early pasture utilization can severely reduce total pasture production (Chapter 2; Noy-Meir, 1978).

Pasture production is region dependent and is given as the mean long-term annual primary production or, more specifically, as herbage dry matter in kg per ha. This is measured at the end of the growing season and represents the total weight of herbage on an ungrazed area of land which was not harvested earlier in the season ('peak biomass'). The base value is determined in the present study on annual pasture vegetation growing on cultivable land, similar to the deep loessial soil in the northern Negev. Pasture productivity on rangeland in the same region is taken as one third of the productivity on cultivated land (Noy-Meir, 1975b; Seligman *et al.*, 1960). Data for other regions are derived from Zaban (1981). Primary production can be increased by adding fertilizer, especially nitrogen (see Chapter 3). However, rangeland with shallow soil, where low moisture availability limits plant growth, usually responds poorly to fertilization and, therefore, fertilizer application on hill range is not considered. Pasture fertilization is also not considered in extensive systems with low target production. They are extensive in the sense that they require low capital and other external inputs. In these systems the available monetary resources will be used to supply the minimum supplementary feed requirements of the flock. The more intensive capital investment required for pasture fertilization is possible when target production is higher. Therefore, pasture fertilization is applied only to intensive and highly intensive systems. However, this general rule may not be applicable to regions where severe deficiency of a mineral element like phosphorus severely reduces production. In such cases correction of the deficiency may be essential, even in extensive systems. Some Australian systems would be a case in point (Donald, 1965; Moore, 1970). The effect of fertilizer application on pasture is negligible, and even negative, when moisture limited pasture production is below 1 ton biomass dry matter per hectare (van Keulen, 1975). In the semi-arid region that the northern Negev represents, yearly biomass production on cultivable land is on average 2.5 tons dry matter per ha. Here average long-term response to fertilizer produces an additional 2.2 tons per ha (Seligman and van Keulen, 1989). The increase in biomass due to fertilizer application is greater in regions where rainfall conditions are better and where both actual and potential production levels are higher. There the amount of fertilizer to be applied is also generally higher, except in particularly fertile soils.

The amount of fertilizer applied can be related to the expected response by a factor that represents the efficiency of nitrogen fertilizer use expressed in kg of biomass produced per kg of nitrogen fertilizer applied. The value used here is 30, assuming a 50 percent recovery of fertilizer nitrogen and a mean seasonal nitrogen concentration of 1.5 percent in the additional biomass (see Chapter 3). The amount of fertilizer nitrogen necessary to attain the target response is then calculated as the ratio between the additional biomass and the fertilizer nitrogen efficiency. The amount of fertilizer per ewe is subsequently calculated as the amount per unit area times the area available per ewe (Section 5.2.5).

All of the total pasture production is not available for utilization by the grazing sheep because some is ungrazable, some is shed as seeds that become trampled into the ground or removed by ants and rodents (Loria and Noy-Meir, 1980; Ofer, 1980), some is removed by wind (de Ridder *et al.*, 1986). The amount of the pasture production available for grazing is taken as three quarters of the peak biomass as defined above (Gutman, 1979).

Pasture quality is divided into three classes:
a. Green season biomass, where quality is high enough to supply all nutritional needs of the lactating sheep. During this stage the nutritional value of the pasture over the whole green season is about 0.77 Scandinavian feed units (FU) per kg dry matter (see below). Pasture utilization is determined by the feed requirements of the ewe for maintenance and production. Total dry matter used is then the total feed requirement during the green season, divided by the feed value of the pasture.
b. High quality dry season biomass soon after the green season when the feed value is on the average 0.55 FU per kg dry matter.
c. Dry season biomass when the feed value drops to 0.33 FU per kg dry matter. During the dry season, the relatively low quality and palatability of the dry pasture limit intake (de Ridder *et al.*, 1986), so that pasture consumption is not determined by requirements only.
During the high quality season, intake is limited to about 1.8 kg dry matter per ewe per day, and during the main dry season to 1.5 kg per ewe per day

(Benjamin *et al.*, 1977). The latter amount is more or less sufficient for maintenance only. For higher target production, supplementation of energy, protein and sometimes minerals is necessary. Transition from higher to lower pasture values in the successive phases is not abrupt, but average values are used in the PSG because the time unit for the pasture cycle is the whole pasture phase rather than a shorter time unit, like days. The actual amounts of feed consumed during these phases depend on the sheep husbandry system (Section 5.2.2).

5.2.2. *Flock*

The basic animal unit for all the input and output calculations in the PSG is the average ewe in the flock. The flock size is determined by labour requirements, which vary with net lambing rate, and availability of labour saving installations like fencing and artificial rearers (see Section 5.2.6).

Flock production in the PSG is limited to two basic products: meat and breeding stock. Wool, which is either coarse carpet wool from Awassi sheep, or medium quality Merino wool, is a minor commodity compared to lamb production, and income more or less covers the shearing costs. Accordingly, neither wool production nor shearing costs are taken into account.

In the present set of systems milk production for sale has not been treated separately, because it requires a different infrastructure and logistic frame of reference. In lamb production flocks, all the milk is generally invested in the lamb. In subsistence systems, some of the milk may be used for human consumption but that would be balanced by lower lamb production. Calculation of input/output relations for a flock devoted mainly to producing milk would need a modified PSG.

Meat production is lamb sold after weaning or fattening and also culled sheep. The maximum flock increase rate is calculated by multiplying the number of female lambs at weaning by a selection fraction after deducting those needed for replacement. The net lambing rate, defined as the number of lambs weaned per ewe in the flock, is a function of breed and management intensity (Table 5–2). The replacement rate is dependent on ewe prolificacy and varies from 15 to 25 percent as prolificacy increases from 0.5 to 2.4. The selection fraction, *i.e.* the fraction of lambs suitable for breeding, is 0.8 for the low and medium prolific Awassi and Merino breeds, and 0.64 for the highly prolific Finn cross.

The liveweight production of the flock is the product of saleweight and the balance of net lambing rate and ewe mortality rate. Saleweight is a target defined for each system (Table 5–2), whereas ewe mortality rate is dependent on prolificacy, increasing from 3 percent to 7 percent per year as prolificacy increases from 0.5 to 2.4. The female lambs available for breeding can be kept to increase breed numbers or sold for mutton. How many will be kept and how many sold is not determined in the PSG but in the technology selection routine (see Chapter 7). The culled sheep fetch about half the lamb price per unit weight. In the calculation of liveweight production, no distinction is made between culled ewes and lambs because the weight of the culls is generally higher than saleweight of the lambs and offsets the lower price per unit weight.

Flock nutrition. The feed requirements in the PSG are expressed in Scandinavian feed units, FU. This is used here in preference to energy, protein and mineral standards because it is a single integrated unit which is sufficient for a regional development study as in the present case. In these terms, green pasture has a value of approximately 0.77 FU per kg dry matter, and pasture in the middle of summer 0.3 to 0.4 FU per kg dry matter (Seligman *et al.*, 1960). The total feed requirement of the flock is derived from requirements for yearlong maintenance, flushing before the breeding season, steaming up before lambing, lactation till weaning, lamb fattening after weaning till sale, replacement hoggets and rams, and milk production if ewes are milked.

Maintenance requirements depend on whether sheep graze or not. Mature sheep require about 0.8 FU per sheep per day when grazing, and about 0.6 FU per sheep per day when enclosed in a holding paddock or corral (Seligman *et al.*, 1981). Sheep are not on pasture during the autumn deferment that is practiced in most of the intensive systems (Table 5-2), and in the highly intensive systems, where grazing is limited to the green season and to the high quality dry season.

Flushing refers to the extra feed that is given to sheep so as to enable them to maintain or attain a body condition suitable for breeding (Coop, 1966b). Flushing continues for about 40 days before the beginning of the breeding season, when good quality feed is given at a rate of 0.5 FU per day. Where the prolificacy of the system is low (less than 1.0 weaned lambs per ewe per year) there is no flushing.

Steaming up before lambing is given to help the ewe overcome pregnancy toxaemia before lambing and to be able to feed the lamb adequately during the critical period soon after lambing (Coop, 1982). The requirement is derived from current practice where 0.3 FU per ewe per day are given to animals with 80 percent net lambing rates (Awassi sheep) and 0.7 FU per ewe per day to those of prolific breeds with 180 percent or higher net lambing rates (Dr. E. Eyal, personal communication). The steaming-up requirements for any system are calculated from these two values by assuming a linear relationship with net lambing rate. Steaming up is continued for 40 days before lambing.

Lactation requirements for the lamb are also dependent on whether the sheep graze or not. A good lactation rate is necessary to allow the lamb to grow out to target weaning weight at a rate of about 0.3 kg per day. At pasture the lactation requirement is about 3.0 FU per kg lamb liveweight increase and 2.5 FU per kg for confined sheep. The total lactation requirement is increased by 5 percent to account for those ewes that lost their lambs before weaning but suckled them for a good part of the lactation period.

Lamb fattening requirements after weaning till sale depend on the feed conversion efficiency of weaned lambs, which decreases as the animals put on weight (Ungar, 1984). The feed requirement per kg liveweight increase can vary between 4.35 and 6.15 FU as liveweight increases from 30 to 45 kg. In extensive systems, where lambs are sold soon after weaning, saleweight equals weaning weight and there is consequently no feed requirement for fattening.

Hogget feed requirement is the feed required to raise weaned lambs to a mean hogget weight of about 50 kg. The number of female lambs retained as hoggets depends on the replacement rate of ewes in the flock. The feed conversion efficiency is taken as a mean 5 FU per kg between weaning and hogget target weight.

Ram feed requirement is on the average 1.0 FU per day or 365 FU per year. As a rule one ram can serve 40 ewes, so the ram feed requirement is added to the ewe unit at a rate of 365/40 FU per ewe per year.

5.2.3. *Feed balance and supplementary feeding*

Green and dry pasture differ so widely in their nutritive value that it is necessary to deal with each separately when constructing a feed balance. On the other hand, the availability of dry pasture depends to a large extent on the utilization of green pasture and so this interrelation must also be taken into account.

Green pasture composed of annual species is, as a rule, a nutritious feed that can supply all the needs of the productive sheep when it is available in sufficient amounts (Eyal *et al.*, 1975; Tadmor *et al.*, 1974). The amount, however, is limited by the length of the effective green season, which in semi-arid regions is relatively short (Noy-Meir, 1975b). Consequently, the green feed will be insufficient for potential animal production and target production levels will accordingly be low. In extensive systems all maintenance requirements are normally obtained from pasture. Supplementary feeding for maintenance is given only in emergencies during severe drought. In more intensive systems, supplementation is not only given for maintenance whenever the flock is off pasture, but for higher production as well.

The feed requirements, over and above maintenance, that can be met by green pasture are, in order of priority: lactation, lamb fattening, steaming up (before lambing), flushing (before breeding). During the effective green season the sheep can satisfy their appetite and nutritional requirements from pasture only. If target weaning weight is attained before the end of the green season, excess pasture can be allocated to other requirements, according to the order of priority as indicated above. If the duration of green pasture season is too short for attaining target weight, the animals must be given supplementary feed to make up the difference. The green season balance (days) is then defined as the length of the effective green season minus the number of days necessary to attain lamb weaning weight. The latter is defined as the difference between weaning weight and birth weight, divided by an assumed daily weight increase of 0.3 kg. To account for lambing distribution and stress days, due to unfavourable weather conditions, an arbitrary value of 15 days is added.

The green pasture balance can be positive or negative, depending on the target weaning weight and on the length of the green season. The balance is converted into feed units by multiplying the number of days of excess or deficient green pasture by the productive feed value of a grazing day over and above maintenance, which must be covered in all cases. In the PSG this is defined as 1.05 FU per day, which is the lactation requirement of a ewe suckling one lamb. The amount is corrected for those ewes that did not lamb and so consumed less pasture. The feed value of the balance is regarded as independent of the net lambing rate because it applies only to the post-weaning phase.

The need for supplementary feeding is derived from the pasture feed balance. If the balance is positive, green pasture is available for maintenance during the green grazing period, for lactation, for steaming up (lambing), flushing (breeding) and maintenance during the deferment period. If the balance is negative, supplementation is required for all these functions.

The supplementation requirement for lamb growth above the target weaning weight of the system is generally met in the feedlot. But, if the green pasture balance is positive, some of the fattening requirement can be met on the pasture and the lambs can be weaned onto a fattening pasture (Ungar, 1984).

The requirement for steaming up which precedes lambing would be next to be met if the green season forage was available above that required for the previously satisfied functions. If the balance is insufficient then steaming-up requirements are met by supplementation.

Similarly, supplementation requirements for flushing are the requirements that can not be met from the remaining green pasture feed balance.

Finally, supplementation for maintenance during grazing deferment could be reduced if the green pasture balance was still positive after meeting the demands of all previous functions. This could occur if the growing season began early and favourable growing conditions enabled the pasture to be ready for grazing before the beginning of lambing. However, this type of situation would be an exception in the study region.

When the green season balance is negative, all the feed deficiencies during the green season must be met by supplementation and all other functions must be fully supplemented. In extensive systems this is generally not necessary as weaning weights are determined by the length of the green season.

The total requirement for supplementation is calculated as the sum of the supplementary feed requirements for the various functions.

5.2.4. *Supplementary feeds*

Having determined the need for supplementary feeding, it now remains to decide what type of feed is necessary. Basically the wide range of feeds available for ruminants can be divided into two main groups: concentrate feeds and roughages. The concentrate feeds are generally high quality, highly digestible feeds, based mainly on grain and oilcake. They can replace much of the quality green pasture if properly balanced nutritionally (see Chapter 4). It is usually one of the most expensive inputs into pastoral systems and so its use is generally kept to the minimum necessary to achieve the production targets. Sown legume pastures like those widely used in Australia, provide high quality feed that can partly replace the expensive concentrates. Roughage feeds are often crop residues (straw) and sometimes special crops grown for hay and silage. The roughages are generally bulky so that even though the cost of straw at source is relatively low, transport costs over large distances can make its use very expensive. Specialized crops for roughage in the semi-arid zone are grown under irrigation (alfalfa) and on dry land (vetch, peas, clover). In such cases the product at source also tends to be expensive and the difference in costs per feed unit between such a roughage and concentrate feed tends to be small. In many regions poultry litter is available as a nitrogen supplement with low energy availability. It is used in combination with straw to provide a balanced roughage for supplementation (Benjamin et al., 1979).

Depending on availability it may be desirable to use more concentrate or more roughage in the supplementation. Two options are defined in the PSG: a high concentrate ration and a low concentrate ration. In the high concentrate option all supplementation needs, except half of the requirement for maintenance during grazing deferment, are met by concentrate. In the low concentrate option, all of the maintenance ration during deferment and one third of the steaming-up ration can be given as roughage. A distinction is further made between 'obligatory' concentrate, that must be supplied by grain, and 'replaceable' concentrate that can be substituted by legume pasture. The latter component can replace the requirements for lactation, steaming up and 80 percent of that for fattening. The required roughage ration is calculated as the difference between the total supplement requirement and that met by concentrate, and the feed requirement derived from pasture is equal to the total feed requirement minus the supplement provided.

5.2.5. *Pasture allocation*

The amount of pasture biomass that can potentially be used is dependent on the length of the green season and the degree to which the dry pasture is used in the different systems. The amount of biomass that can potentially be consumed during the green season is equal to the length of the effective green season times the feed requirements for maintenance and lactation (Section 5.2.2), converted into biomass. For conversion the feed value of green pasture is taken as an average value of 1.3 kg dry matter per FU for the whole green season.

The intake of dry pasture is dependent on the digestibility of the pasture rather than on animal requirement, and dry matter intake is generally limited to 1.5 kg per ewe per day for most of the dry season and 1.8 kg per ewe per day during the early high quality phase (de Ridder et al., 1986). In extensive systems and in intensive systems that are implemented in regions where the dry pasture season is very long, the potential dry pasture utilization in kg dry matter per ewe is equal to 1.5 kg per day times the length of the dry season in days; in highly intensive systems and other 'green season grazing systems' it is equal to 1.8 kg per day times the length of the high quality dry pasture phase, which is taken as half the length of the green season (Section 5.2.1). The total potential biomass utilization is the sum of potential green season utilization and potential dry season utilization.

It is now possible to calculate the area of pasture per ewe for each system. The mean peak biomass on the pasture per unit area and its utilizable fraction have been defined above (Section 5.2.1). The area needed per ewe is equal to the total pasture dry matter requirement per ewe divided by the utilizable biomass per unit area.

This approach assumes that the main pasture production is fully utilized and ignores over- or under-utilization. This is an oversimplification in extensive systems, but that is partly taken care of by assuming lower target production. In supplemented systems this may not be too serious an error considering the flexibility afforded by supplementation that can to some degree be increased or decreased according to annual fluctuations around the long-term mean.

5.2.6. *Labour requirements*

The work needed in a flock of sheep maintained mainly for lamb production includes herding on pasture or on crop aftermath, care for livestock, maintenance of equipment, fences and other structures and management. Herding, especially on fenced land, is not closely related to the size of the flock. Where flocks are kept on pasture continuously throughout the day and night, herding requirements are relatively low and involve mainly regular inspection to take care of unexpected events. In most mediterranean regions sheep are herded daily to and from a central corral, even when fences are available, because of predation and theft, both of which are serious and sometimes incur crippling costs in lamb raising flocks.

The most intensive labour activities are those associated with care of animals. Even in extensive wool flocks in Australia the labour requirements for shearing often exceed the capacity of the flock owner, who must call in contract shearers to help him out (Moore, 1970). In semi-arid regions the periods of intensive work in fat-lamb flocks are not concentrated into relatively short bursts of high activity, but involve a more diffuse work demand that includes special attention during breeding, intensive care during the lambing season and daily feeding chores during lamb fattening and during periods of grazing deferment. The amount of work needed increases with the number of sheep that lamb and with the number of lambs to be reared. The more prolific the ewe the more care is needed during lambing and in the raising of lambs. The ewes also need more care as the complications that result from large litters are greater than from single-lamb births (Coop, 1982; Spedding, 1970). As the care of the young growing animals is the main function that determines the size of the flock that can be handled by one man, and as care needed increases with the mean litter size at lambing, the labour requirement of the flock can be related to ewe prolificacy. It has been difficult to obtain hard data on this relationship and those used in Table 5–3 are an estimate that

Table 5–3.
System parameters dependent on ewe prolificacy.

Ewe prolificacy	Size of flock				Ewes that lambed	Shed area	Vet. costs
(NETLMR) weaned lambs	(HERDSZ) number of ewes				(ETL) fraction	(BSTRUC) m² ewe⁻¹	(VETCOS) $ ewe⁻¹ yr⁻¹
	with fences		without fences				
	ewes	lambs	ewes	lambs			
0.50	500	250	300	150	.55	0.0	1.0
0.75	400	300	250	188	.80	0.1	2.0
0.90	350	315	220	198	.82	0.7	4.0
1.50	250	375	180	270	.87	1.0	6.0
1.80	200	360	150	270	.90	2.0	7.0
2.40	150	360	125	300	.95	2.5	10.0

represents a consensus of opinions between sheep extension officers and research workers in Israel.

Labour input in flocks with low prolificacy is relatively heavy because the herding component is not closely related to the size of the flock. In highly intensive systems the smaller number of lambs weaned per man-year is due to the fact that high weaning rates are achieved by multiple lambing, which requires more care, including labour for administration of hormones to initiate and synchronise oestrus in off-season breeding (see Chapter 4). As these systems are labour intensive, it has been suggested that labour saving artificial rearers can reduce labour requirements and also improve lamb survival. Here too, hard data are not readily available, but it is suggested that the use of an artificial rearer can increase overall handling capacity by about 30 percent. This represents a fairly high level of technology and in the PSG is applied as an option in highly intensive systems only.

Where animals are kept under full confinement with no pasture, no herding is involved and the labour requirement is reduced. In the PSG a 20 percent increase in flock size is assumed over what can be handled by one man in systems on pasture with fences. With an artificial rearer the flock size would be limited mainly by the handling capacity of one man with the artificial rearer. This is taken here as 200 lambs (Dr. E. Eyal, personal communication; see also 4.5.3).

The labour requirement per ewe is now simply defined as the reciprocal of flock size, the latter being the number of ewes that can be handled by one man during a year. The man-year includes weekends and holidays and amounts to about 250 to 280 full work-days per year. In extensive systems without fencing the man-year would be closer to 360 work-days, with some work being done by members of the family.

5.2.7. *Capital requirements*

Capital investments include overnight corrals, lambing and fattening sheds, fences and watering points on the pasture and equipment, including tractor, spraying equipment and miscellaneous items. In highly intensive systems this would also include artificial rearer units and the accompanying building structures.

The number of overnight corrals required depends very much on the size and geographical distribution of the pasture and on the security of the area with regard to theft and predation. All systems need at least one corral; an additional one suitably placed to serve more distant pastures, is often the minimum requirement. Consequently, two corrals per flock have been postulated. The cost of such a unit is estimated at $ 2 per ewe.

The cost of lambing and fattening sheds and other structures depends very much on lambing rate and on system intensity. The area required per ewe, according to local sheep husbandry extension standards, is given in Table 5–3. The cost of these structures is estimated at $ 20 per m².

The length of necessary fencing is dependent on the area and shape of the pasture and on the number of paddocks or subdivisions. A more or less square area is assumed, so that the circumference in meters is 4 * F * SQRT(S), where S is the area of pasture in m², and F is a factor greater than unity, accounting for some irregularity in the shape of the pasture. The dividing fences are each the length of one side of the square pasture. The number of dividing fences depends on the number of paddocks in the pasture. A minimum of two dividing fences, making four paddocks, is assumed to allow for flock and pasture management. The total length of fence is then obtained from the above equation by substituting area per ewe times flock size for S and adding two dividing fences to the circumference; F has been set at 1.2. This calculation assumes that there are no neighbours to share the cost of the circumference

Table 5–4.
Minimum number of dividing fences necessary for a given number of paddocks.

Number of paddocks (P)	Number of subdividing fences (NFENCE)
1	0
2	1
4	2
6	3
9	4
12	5
16	6

P = (NFENCE/2+1)², holds for even values and is a close approximation for odd values of NFENCE.

fences. Where there are neighbours, the length of circumference fencing debited to pasture should be accordingly less. On the other hand fencing requirements would increase if more than four grazing paddocks are required. The number of paddocks that can be created with a given number of subdividing fences is variable, but the maximum is given in Table 5–4. Cost of fencing is taken as $ 1,500 per km.

Water supply can vary from spring or river water available year-round at no cost, to piped water that involves both cost of investment in pipes and watering points as well as the cost of the water itself. In the PSG it is assumed that the water is not free and that installations are necessary for using it. As a standard it is assumed that at least one watering point is required on every 500 ha. The number of watering points per ewe is then equal to the area per ewe divided by that value. Further it is assumed that each such watering point requires 5 km of piping and a watering trough for each 200 ewes. The total capital cost of water per ewe is then the total required pipe length times the price per unit of length plus the price of a trough, divided by 200. The cost of piping is set at $ 1,000 per km and that of a trough at $ 150.

If there is no pasture use, like in some highly intensive completely confined systems, both the cost for fencing and water are, of course, zero.

Equipment costs, including tractor, shearing and spraying equipment and miscellaneous items like small generators, water pumps and tools, are lumped and taken as a cost per ewe for a standard flock size. The cost of equipment per unit flock size of 500 ewes is around $ 10,000 or $ 20 per ewe. The cost of equipment per ewe is thus $ 10,000 divided by the size of the flock.

If the pasture is fenced, then the whole cost of equipment is charged. If not, the whole cost is charged only to relatively intensive systems with a net lambing rate of 1.1 or more. For less intensive systems without fencing the equipment cost is reduced to half, to account for the fact that capital costs in general will be much lower in such systems.

An artificial rearer that can effectively save labour, costs about $ 10,000 per unit and can serve lambs from 200 prolific ewes. The cost per ewe is then the ratio of these two numbers.

5.2.8. Running (or current) costs

This item includes costs of supplementary feed, veterinary costs, water, and miscellaneous materials. Supplementary feed is divided into concentrate feed that is bought generally from outside the region and roughage that, as a rule, is produced on the farm or traded only within the region (Section 5.2.4).

Water requirement can be estimated from the total feed requirement and amounts to about 6 litres per feed unit (Benjamin et al., 1977). Water is also needed in the lambing sheds; approximately 1 m³ per m² of built structure. Water is saved during the green season when the water requirement is met by the green pasture. The annual water requirement is obtained from the total feed requirement and the built structures present, corrected for the length of the effective green season. The cost of water is very variable from place to place, but is taken here as $ 0.20 per m³.

Annual veterinary costs include professional care, medicines, hormones for synchronizing oestrus, and pesticides. The veterinary costs increase with prolificacy as the larger number of lambs, the cost of hormones and the heavier production load on the ewes all expose the flock to greater health hazards. The relationship in Table 5-3 is taken from local extension practice (Dr. E. Eyal, personal communication).

All costs are given in US dollars as calculated in 1985.

5.3. Cropping systems

Three cropping systems are defined: (i) continuous wheat, where wheat is grown for four to five consecutive years and then fallowed for one year, (ii) a wheat/fallow rotation, where wheat is grown and the land is fallowed on alternate years, and (iii) a wheat/legume rotation, assumed to cover equal periods, say three years of legumes followed by three years of wheat.

Utilization of wheat straw is defined as a separate system, dependent on the amount of wheat grown. The yields of wheat, straw and legume pastures are dependent on the climate, mainly rainfall; zone and mean yields are given in Table 5-5.

Capital investments for wheat cultivation are separated into machinery for cultivation and sowing, set at $ 476 per ha, and for harvesting at $ 255 per ha. Sowing and harvesting capital is reduced in the more arid zones by a factor to account for more extensive cultivation systems. This factor varies from 0.33 in the Western Negev (zone 2) through to 1.00 in the central coastal plain zone. The current costs for wheat cultivation and straw baling are taken as $ 115 and $ 10 per ha, respectively (Zaban, 1981).

Table 5-5.
Mean yield of field crops in different zones in the semi-arid belt in Israel (kg ha⁻¹ year⁻¹)[1].

Zone	Continuous Grain	wheat straw	Wheat-fallow Grain[2]	Wheat-fallow Straw[2]	Legume
1. Negev highlands	0	0	0	0	350
2. Western Negev	360	750	1,175	1,240	1,800
3. Beersheba plain	1,032	1,293	1,575	1,900	2,400
4. Northern Negev	1,520	2,015	2,250	2,900	3,000
5. South coastal plain	2,476	3,190	3,540	4,425	3,900
6. Central coastal plain	3,750	4,500	5,175	6,090	4,800

[1] Source: Zaban, 1981
[2] Mean yields for cropped area, excluding the fallow area.

For the wheat/legume rotation the investment costs for sowing and cultivation machinery and equipment are equivalent to those for wheat at $ 476 per ha. Harvesting costs do not apply to the legumes, while the current costs for land preparation, sowing and fertilizer are set at $ 115 per ha.

5.4. PSG activity table

The linear programming (LP) routine, used for the technology selection phase (Chapter 7), requires that all the input/output values that describe each of the systems are arranged in an appropriate matrix format, called an 'activity table'. For each system ('activity') the PSG calculates a vector composed of input/output values ('technical coefficients'). This set of vectors is then arranged in a table that constitutes the 'A-matrix' of the LP tableau (Chapter 7, Fig. 7–1). The inputs for each activity are defined as requirements and include:
- area of rangeland or cultivable land;
- length of fencing on rangeland or cultivable land;
- value of other capital investment;
- artificial rearers needed in intensive systems;
- roughage and concentrate feed requirements;
- labour requirements;
- the number of hoggets per ewe (Awassi, Improved Awassi, Merino, Finn cross) available each year for flock increase.

The output is the amount of lamb and mutton produced per ewe. Straw that is produced within the region as a by-product of small grain cultivation is not sold outside the region. It can be a constraint that limits the size of the flock, especially if the region tends to specialize in very intensive systems without pasture.

The basic capital investment in pasture on cultivable land or rangeland includes water, equipment, corrals and lambing sheds. Fencing and artificial rearers are listed separately as optional investments. The basic capital for herds on both cultivable and rangeland includes the costs per ewe of built structures (lambing sheds), corrals, equipment and watering points, respectively. The cost of fencing is the product of the length of fencing (km per ewe) and cost of fencing per unit length ($ per km).

The number of hoggets available for flock increase is divided into the different breeds and can be a strong constraint on expansion of desirable systems when there is only a small initial number of ewes of the desired breed. The maximum number of hoggets available for breeding each year is always the maximum flock increase as defined in Section 5.3.2, but the actual number and breed will depend on system selection.

5.5. **Input and output values calculated by the PSG**

Examples of the range of values calculated by the PSG for a selected set of variables are presented in Figures 5–1 to 5–6. The output values illustrated in the diagrams cover the main aspects of pasture productivity, animal performance, feed requirements, production and capital input. Each figure consists of 18 groups of vertical columns, that represent the 18 basic pastoral systems (Table 5–2). Each one of these systems can be implemented under different land type, fencing and feeding conditions (Table 5–1). Not all combinations are realistic (see Section 5.1.2) so that the eight defined in Table 5–1 can be condensed into the following six categories:

I   – cultivable land, unfenced, low concentrate option;
II – cultivable land, unfenced, high concentrate option;
III – cultivable land, fenced, low concentrate option;
IV – cultivable land, fenced, high concentrate option;
V   – systems 1–11: rangeland, unfenced, low concentrate option;
      systems 12–18: no pasture, low concentrate option;
VI – systems 1–11: rangeland, fenced, high concentrate option;
      systems 12–18: no pasture, high concentrate option.

When the 18 basic pastoral systems are varied according to these six categories, they add up to 108 systems (or 'activities'). In the Figures 5–1 up to and including Figure 5–5 the variables illustrated are not dependent on either fencing or on feeding system. Consequently, for each of the 18 basic systems, the values of the variables are identical within the group of cultivable land categories I to IV and within the group of rangeland/confinement categories V and VI. That is why each group is represented by one column only for each of the 18 basic systems. In Figure 5–6, the illustrated variables are dependent on fencing or feeding system and so each of the six categories must be represented separately.

Pasture herbage production and allocation. Herbage production from pasture is a regional characteristic, but varies with land type (cultivable or rangeland) and fertilizer application (Fig. 5–1a). The pasture herbage requirement per ewe depends on system definition and varies from no pasture through to almost full dependence on pasture (Fig. 5–1b). As a consequence, pasture allocation (ha per ewe) also varies over a wide range from no pasture to almost 1 ha per ewe (Fig. 5–1c). The contrast between rangeland and cultivable land allocation is particularly striking.

The grazing season. The division of the year into grazing seasons depends not only on the climatic pattern, but also on land type, grazing management, pasture fertilization and system intensity (Fig. 5–2). Deferment of grazing to allow pasture to develop to the stage when growth rate exceeds consumption rate, is not prescribed in the most extensive systems and is between 80 and 100 days in the more intensive systems (Fig. 5–2a). The length of the green season

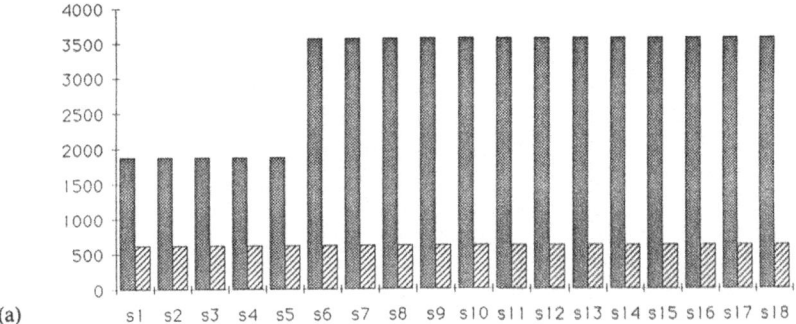

(a)

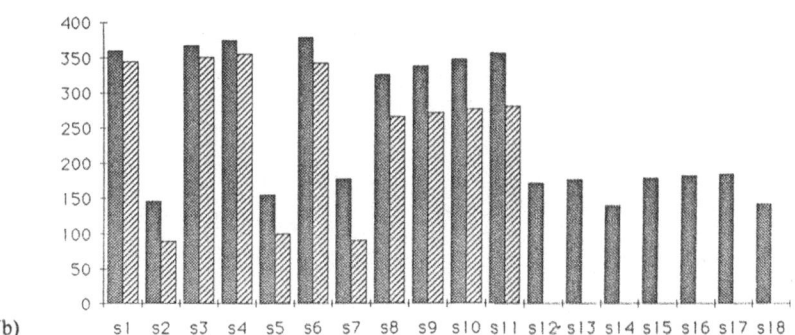

(b)

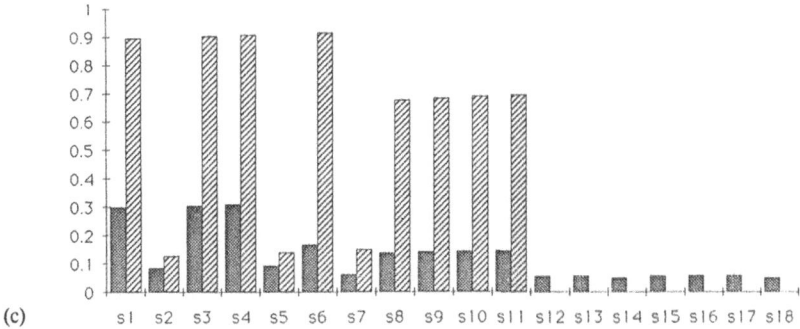

(c)

Fig. 5-1. Pasture variables.
a. Pasture herbage dry matter available for grazing, kg ha^{-1}
b. Annual pasture-herbage requirement, Feed Units (FU) ewe^{-1}
c. Pasture allocation, ha ewe^{-1}

124

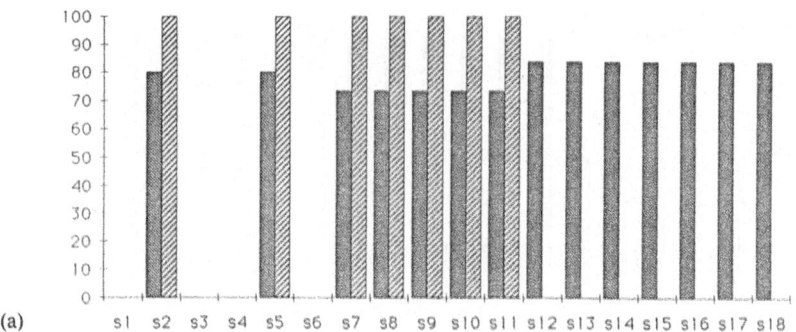

(a)

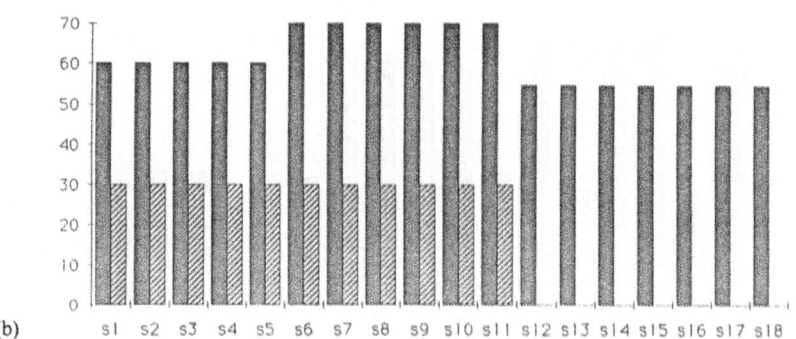

(b)

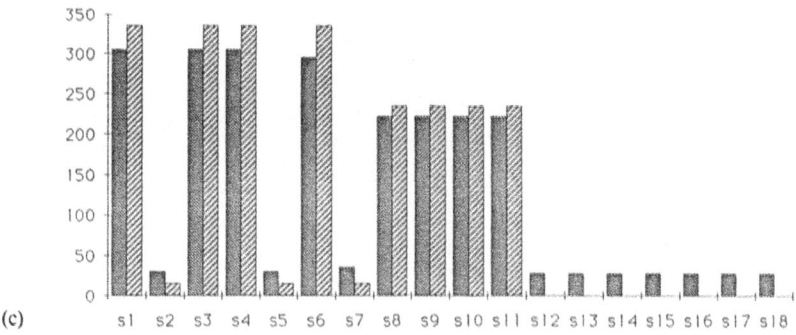

(c)

Fig. 5-2. Grazing period variables (days).
a. Early season grazing deferment
b. Duration of green season
c. Duration of dry pasture grazing

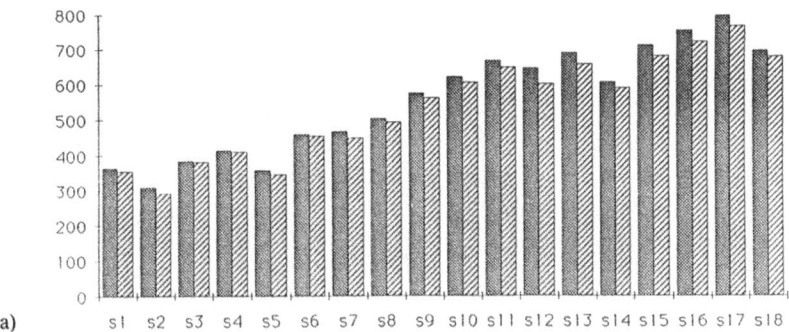

(a)

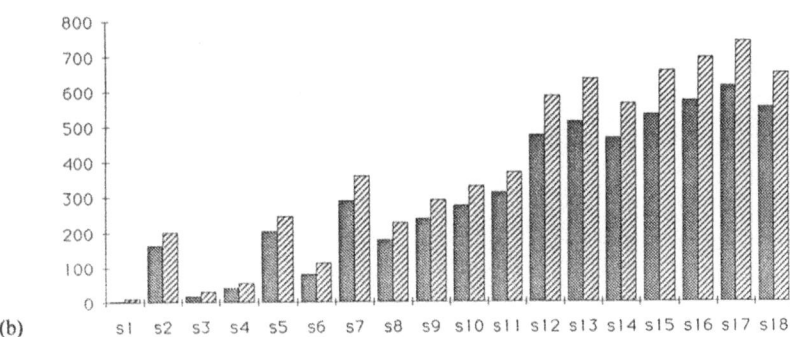

(b)

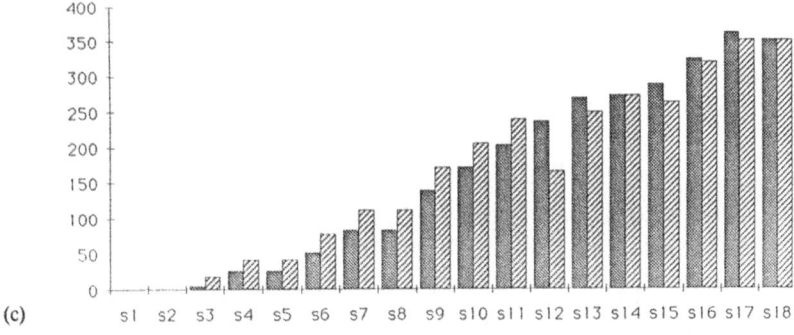

(c)

Fig. 5-3.
Feed requirements of the flock (FU ewe^{-1} year^{-1}).
a. Total feed requirements
b. Total supplementary feed requirements
c. Feed requirements for fattening

126

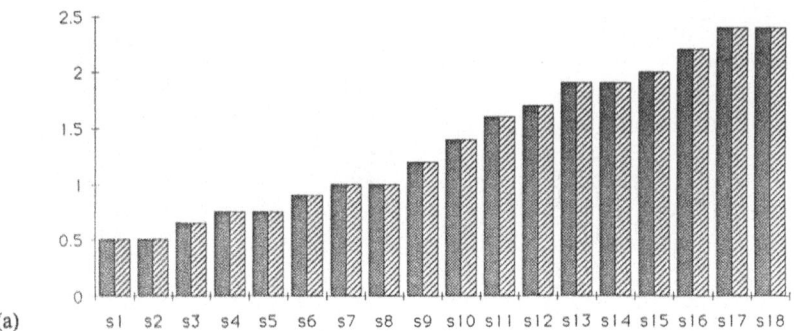

(a)

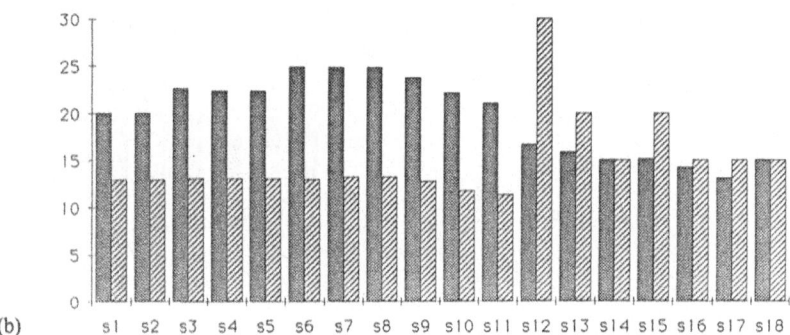

(b)

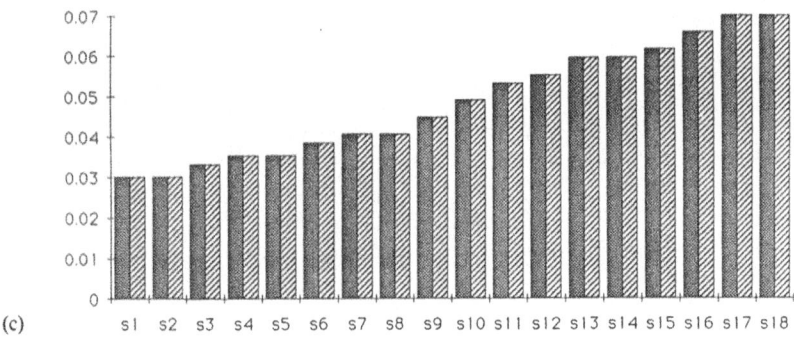

(c)

Fig. 5-4. Sheep performance variables.
a. Net lambing rate, lambs ewe^{-1} year^{-1}
b. Lamb weight at weaning, kg lamb^{-1}
c. Ewe mortality rate, ewes ewe^{-1} year^{-1}

(Fig. 5–2b) is determined mainly by the duration of deferment and by land type. It is relatively short on the rangeland because of shallow soil with low water-holding capacity. The duration of dry pasture grazing is mainly dependent on system intensification level, and is very short in the more intensive systems (Fig. 5–2c). In some of the systems the animals are completely confined and use no pasture at all.

Feed requirements. Total feed requirements per ewe increase with system intensity (Fig. 5–3a), but much of this is supplied by supplementary feed (Fig. 5–3b) and not by pasture. The main component of the supplementary feed is the requirement for lamb fattening (Fig. 5–3c). Differences within the system groups are due to differences in pasture utilization and land type.

Animal performance. The net lambing rate per ewe by definition increases with system intensity (Fig. 5–4a), but lamb weaning weight varies widely with land

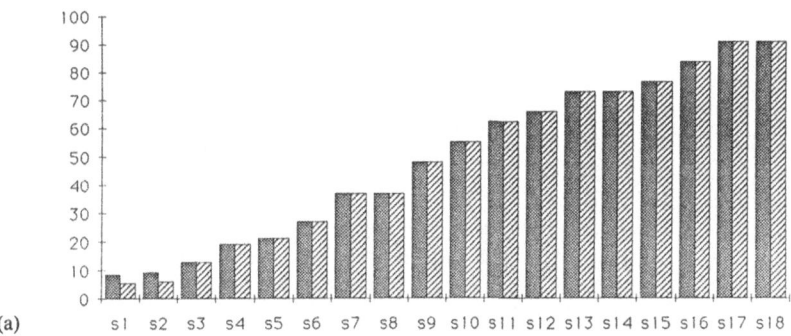

(a)

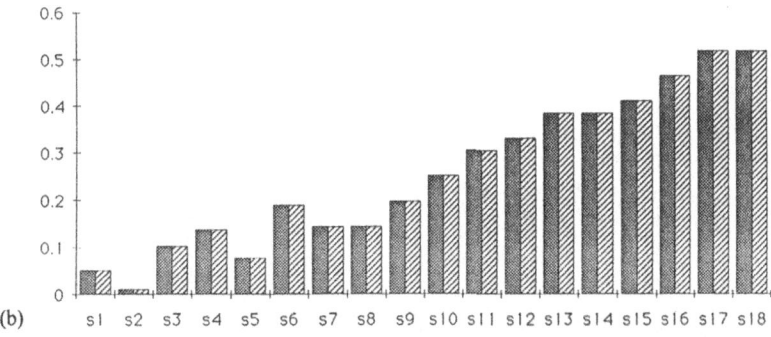

(b)

Fig. 5-5. Flock production variables.
a. Maximum meat output, kg ewe^{-1} year^{-1}
b. Potential herd increase rate, hoggets ewe^{-1} year^{-1}

128

type (more precisely, pasture type), grazing management and system intensity. The weaning weights range between just over 10 kg per lamb till almost 30 kg per lamb (Fig. 5–4b). Ewe mortality increases with system intensity because of multiple births, multiple lambings, danger of pregnancy toxaemia, etc. (Fig. 5–4c).

Potential production and herd increase. The maximum annual potential meat output per ewe is based on the assumption that all the hoggets over and above those required for ewe replacement, are sold and are not raised (Fig. 5–5a). On the other hand, potential herd increase rate assumes that all hoggets suitable for breeding are indeed raised (Fig. 5–5b). In extensive systems the maximum flock increase rate is less than 10 percent per year, while in the highly intensive flocks, the potential animal increase rate can be more than 50 percent because of the high lambing rate in these systems.

Concentrate and capital investments. The concentrate feed requirements show a distinct increase with system intensity. It stands to reason that concentrate use

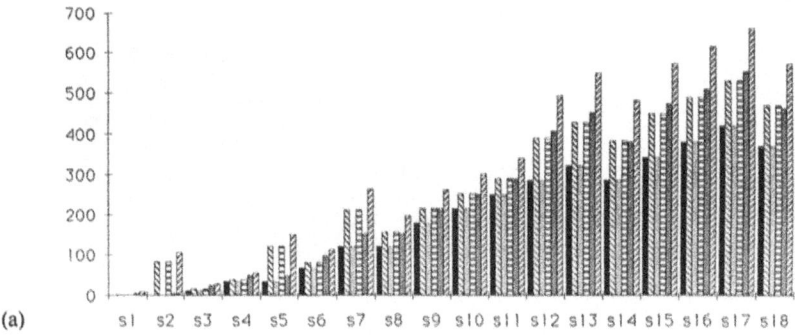

(a)

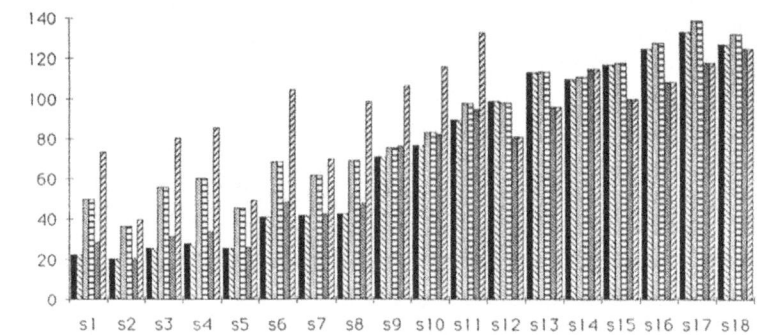

(b)

Fig. 5-6. Concentrate requirements and capital investments.
a. Concentrate requirements, FU ewe⁻¹
b. Capital investments, $ ewe⁻¹.

is highest in the feedlot group (VI), followed by the two high concentrate groups (II and IV). These differences are very pronounced, especially where the Awassi sheep are used, because very extensive systems that use very little concentrate are described only in the case of the Awassi.

The money required for capital investments per ewe also varies with system intensity (Fig. 5–6), but it should be noted that some of the higher capital requirements per ewe are on rangeland where fencing of large areas with low productivity can be a heavy fixed investment per ewe. In the highly intensive systems, the artificial lamb-rearers add significantly to capital costs.

These figures illustrate some system characteristics and can provide a quick comparative reference when analyzing the solutions presented in Chapters 6 and 7 where different systems are selected for different development pathways.

Appendix

Equations used in the PSG for calculating inputs and outputs of pastoral systems.

I. *Equations applied in Section 5.2.1*

TG	$= f$ (region, soil type) (Table 5–6)
TX	$= 120-0.67*TG$
TQ	$= 0.5*TG$
TD	$= f$ (system intensity), *i.e.*

$$TDe = 365-TG$$
$$TDi = 365-TG-TX$$
$$TDh = 0$$

where TG, TX, TQ, TD indicate duration (days) of green season, early season deferment, high quality dry, and dry season respectively; and e, i, h indicate extensive, intensive and highly intensive systems, respectively.

PEAKB	$= f$ (region, soil type) (Table 5–6)
PEAKBM	$= PEAKB + FERTBM$
FERTBM	$= 1.5*(PEAKB-1,000), FERTBM \geq 0$
AFERT	$= FERTBM/FERTEF$
AVPAST	$= 0.75*PEAKBM$

where PEAKB, FERTBM, AVPAST, AFERT, FERTEF indicate peak biomass, biomass increment due to fertilizer application, utilizable biomass (kg dry matter per ha), amount of fertilizer applied (kg per ha), and fertilizer efficiency, respectively.

II. *Equations applied in Section 5.2.2*

POTINC	$= (0.5*NETLMR-REPLRT)*SELFAC$
MOINCH	$= SALEWT*(NETLMR-POTINC-EWEMRT)$
REPLRT	$= (NETLMR-0.5)*0.10/1.9+0.15$
EWEMRT	$= (NETLMR-0.5)*0.04/1.9+0.03$

where POTINC, NETLMR, REPLRT, EWEMRT, SELFAC, MOINCH indicate potential herd increase rate, net lambing rate, ewe replacement and ewe mortality rates (all rates per ewe per year),

Table 5–6.

Mean duration of green season (d) and peak pasture biomass (kg ha⁻¹ year⁻¹) in different zones in the semi-arid belt in Israel [1].

Zone	Duration of green season (TG)	Peak pasture biomass (PEAKBM)
1. Negev highlands	30	400
2. Western Negev	40	1,500
3. Beersheba plain	50	2,000
4. Northern Negev	60	2,500
5. South coastal plain	75	3,250
6. Central coastal plain	90	4,000

[1] Sources: Noy-Meir, 1975; van Keulen, 1975; Seligman and van Keulen, 1989.

selection factor, and maximum lamb output of increasing herd (kg ewe⁻¹), respectively.

MAINTR $= f$ (system intensity), *i.e.*
MAINTRe $= 365*0.8$
MAINTRi $= 365*0.8-TX*0.2$
MAINTRh $= 365*0.8-(365-TG-TQ)*0.2$
MAINTRc $= 365*0.6$

(e, i, h, c: extensive, intensive, highly intensive, confinement without pasture)

FLUSHR $= \begin{cases} 0.4*40, & \text{NETLMR}>1.0 \\ 0., & \text{NETLMR} \le 1.0 \end{cases}$

STEAMR $= 0.4*(\text{NETLMR}-0.05)*40$

LACTR $= f$ (system intensity), *i.e.*
LACTRe = LACTRi = (WEANWT-BIRWT)*NETLMR*1.05*3
LACTRh = (WEANWT-BIRWT)*NETLMR*1.05*2.5

FATNR $= (\text{SALEWT-WEANWT})*\text{NETLMR}*(0.75 +$
$0.12*0.5*(\text{SALEWT} + \text{WEANWT}))$

HGTRQ $= \text{REPLRT}*(50-\text{WEANWT})*5$

RAMRQ $= 365/40$

TOTREQ $= \text{MAINTRc} + \text{FLUSHR} + \text{STEAMR} + \text{LACTR} +$
$\text{FATNR} + \text{HGTRQ} + \text{RAMRQ}$

where MAINTR, FLUSHR, STEAMR, LACTR, FATNR, HGTRQ, RAMRQ, TOTREQ indicate feed requirements (Feed Units per ewe per year) for ewe maintenance, flushing, steaming up, lactation, lamb fattening, hogget raising, ram requirements, and annual total, respectively.

III. *Equations applied in Section 5.2.3*

BALGRN $= \text{TG-GPN}$
GPN $= (\text{WEANWT-BIRWT})/0.3 + 15$
BALGFU $= \text{BALGRN}*1.05 + \text{TG}*(1-\text{ETL})*0.5$
ETL $= f$ (prolificacy) (Table 5–3)

where BALGRN, GPN, BALGFU, ETL indicate green season balance (days), period between birth and weaning (days), green season balance (FU), and proportion of ewes that lambed, respectively.

SFATN	= FATNR-BALGFU,	SFATN $\geq$ 0
SSTEAM	= STEAMR-(BALGFU-SFATN),	SSTEAM $\geq$ 0
	(BALGFU-SFATN) $\geq$ 0	
SFLUSH	= FLUSHR-(BALGFU-SFATN-SSTEAM),	SFLUSH $\geq$ 0
	(BALGFU-SFATN-SSTEAM) $\geq$ 0	
SMAINT	= PDFMQ-(BALGFU-SFATN-SSTEAM-	
	SFLUSH)	SMAINT $\geq$ 0
	(BALGFU-SFATN-SSTEAM-SFLUSH) $\geq$ 0	
PDFMQ	= PDFM*0.6	
PDFM	= 365-TG-TD	
SLACT	= min (LACTR, – BALGFU)	

$$SHGTR = \begin{cases} HGTRQ * 0.5, & NETLMR > 0.9 \\ 0, & NETLMR \leq 0.9 \end{cases}$$

$$SRAMR = \begin{cases} RAMRQ * 0.25, & NETLMR > 0.9 \\ 0, & NETLMR \leq 0.9 \end{cases}$$

$$TOTSUP = \begin{cases} SLACT + SFATN + SSTEAM + SFLUSH + \\ SMAINT + SHGTR + \\ SRAMR, & GRAZE = +1 \\ \\ TOTREQ, & GRAZE = -1 \end{cases}$$

where SFATN, SSTEAM, SFLUSH, SMAINT, SLACT, SHGTR, SRAMR, TOTSUP indicate supplementary feed requirements (FU per ewe per year) for lamb fattening, steaming up, flushing, ewe maintenance, lactation, hogget raising, ram requirements, and annual total respectively.

IV. *Equations applied in Section 5.2.4*

$$LRUF = \begin{cases} 0.5 * SMAINT, & GRAZE = +1 \\ \\ 0.5 * MAINTRc, & GRAZE = -1 \end{cases}$$

$$HRUF = \begin{cases} SMAINT + 0.33*SSTEAM + 0.5* \\ (SHGRT + SRAMR), & GRAZE = +1 \\ \\ MAINTRc + 0.33*STEAMR + 0.5* \\ (HGTRQ + RAMRQ), & GRAZE = +1 \end{cases}$$

HICONC	= TOTSUP-LRUF	
LOCONC	= TOTSUP-HRUF	

$$CONC = \begin{cases} HICONC, & MINRUF = 0 \\ \\ LOCONC, & MINRUF = 1 \end{cases}$$

CONCR	= SLACT + SFLUSH + 0.8*SFATN	
CONCOB	= (CONC-CONCR),	CONCOB > 0
ROUGH	= TOTSUP-CONC	
PASTR	= TOTREQ-TOTSUP	

where LRUF, HRUF are roughage requirements (FU per ewe per year) and HICONC, LOCONC are concentrate requirements (same units) under minimum roughage (MINRUF = 1) or minimum concentrate (MINRUF = 0) feeding regimes, respectively; CONCR, CONCOB are concentrate allocations that can be replaced by high quality legume pasture, and 'obligatory' concentrate that cannot be replaced; ROUGH, PASTR are roughage supplementation and pasture requirements after accounting for concentrate allocation, CONC.

V. *Equations applied in Section 5.2.5*

PUG	=	$TG*(0.8 + 1.05*NETLMR)*GPV$
PUD	=	f (system intensity), *i.e.*
		$PUDe = PUDi = 1.5*TD$
		$PUDh = 1.8*TD$
PUT	=	$PUG + PUD$
AREAPE	=	$PUT/AVPAST$
PASPR	=	$PASTR/AREAPE$

where PUG, PUD, PUT, GPV indicate potential utilization of green, dry and total pasture (kg dry pasture biomass per ewe per year), and green pasture value (kg dry biomass per FU), respectively; and
AREAPE, PASPR indicate allocation of pasture (ha per ewe) and pasture productivity (FU per ha per year), respectively.

VI. *Equations applied in Section 5.2.6*

HERDSZ	=	f (prolificacy) (Table 5–3)
LABPEW	=	$1/HERDSZ$

where HERDSZ, LABPEW indicate herd size (number of ewes), and labour requirements (man-years per ewe), respectively.

VII. *Equations applied in Section 5.2.7*

CORRAL	=	$2/HERDSZ$
NFENCE	=	f (number of paddocks) (Table 5–4)
TOTFEN	=	$SQRT(AREAPE*HERDSZ*F) * (4 + NFENCE)*0.1$
FENCEW	=	$TOTFEN/HERDSZ$
WATPEW	=	$AREAPE/500$
CWPEW	=	$KMPIPE*WATPEW*COSTPP + TROUGH/200$
EQUIP	=	$EQUC*500/HERDSZ$
HITEC	=	$CARU/NEWPAR$

where CORRAL indicates number of corrals per ewe;
NFENCE, TOTFEN, FENCEW indicate number of dividing fences, total length of fences (km), and fence length per ewe, respectively;
WATPEW, CWPEW, KMPIPE, COSTPP, TROUGH indicate number of water points per ewe, cost of a water point ($ per ewe), length of pipe to water point (km), cost of pipe ($ per km), cost of permanent water trough ($);
EQUIP, EQUC indicate cost of equipment ($ per ewe), and cost of a standard set of equipment, including vehicles ($);
HITEC, CARU, NEWPAR indicate cost of artificial rearer ($ per ewe), cost of an artificial rearer unit with installations ($), number of ewes served per artificial rearer unit, respectively.

VIII. *Equations applied in Section 5.2.8*

WATREQ	=	$TOTREQ*0.006 + BSTRUC*1-TG*0.004.$

where WATREC, TOTREQ, BSTRUC (see Table 5–3), TG indicate water requirement (m^3 per ewe per year), total feed requirement (FU per ewe per year), built structures (m^2 per ewe), duration of green season (days), respectively.

6. Management of agro-pastoral systems at the farm level

E.D. UNGAR

6.1. **Introduction**

This chapter and the following one are concerned with the integration of the 108 different options of the PSG as described before, but with different objectives. Here the management of systems at a farm level is dealt with, while Chapter 7 is concerned with the development of a region.

The farmer's goals include improving or stabilizing income, reducing risk and reducing labour requirements. The degree of emphasis placed on each goal differs between farmers. However, an almost universal goal is that of improving income, and this is often the sole criterion for decisions.

The purpose of the present chapter is to investigate the economic feasibility and technical robustness at the farm level of some of the systems that were defined in the preceding chapter. Those systems cover a wide range of technologies from subsistence on local resources to intensive use of imported inputs. This chapter focuses on an intermediate level of intensification where local land and pasture resources are the foundation of the systems, but where there is free intercourse with markets and sources of inputs. They are based on land that can be cultivated and used for grain cropping under rainfed conditions, or used as pasture. They also include a flock of sheep of intermediate fertility that is maintained mainly for fat lamb production. They have access to supplementary feed and the necessary veterinary and logistic support.

6.1.1. *Objectives*

The linkage between regional planning and farm level management raises certain questions. Is a farming system that is deemed viable on the basis of its mean annual input-output relations also viable when the year by year variation is examined? Can the broad principles on which agro-pastoral systems are based be translated into workable management rules? For example, the idea that green wheat can be grazed as an alternative to grain production in drought years seems simple on a yearly time scale. But in reality the manager cannot wait to see what the grain yield will be and how bad a drought year it will be before making the decision to graze. How efficient are these systems when variability

133

Th. Alberda et al. (eds.), Food from Dry Lands, 133–158.
© 1992 *Kluwer Academic Publishers.*

and uncertainty are explicitly taken into account? How stable is the technical efficiency of an agro-pastoral system to management options and the price regime?

For purposes of management at the farm level, a primary objective is to devise a set of optimum or near-optimum management rules. Relevant questions at this level of resolution include: What are the important management decisions? How should these decisions be treated? How sensitive is system performance to management?

The above questions are addressed by using a management model of an agro-pastoral system that explicitly considers the within-season physical and biological dynamics of the system.

6.1.2. *The agro-pastoral system*

The agro-pastoral system referred to in this chapter is a farm unit which has four basic features:
- A grain production component, assumed here to be wheat. The grain is produced primarily for sale.
- A pastoral component that includes both pasture proper and grazing of wheat or wheat residues.
- A flock of sheep that derives a significant portion of its feed requirements by grazing the pastoral component, and produces lamb meat for sale.
- A management system in which grain and meat production is integrated, meaning that the crop component can be utilized for grain and/or grazing by the sheep flock according to whole-system considerations.

Management decisions are not inherent features of the physical and biological components of the system. Management options are man-made, and it is fair to assume that not all useful options have been devised. Management options are also a function of the environment in which the system operates, and no claim is made for global generality or applicability. There is assumed to be a developed transport, marketing and veterinary infrastructure, a competent level of managerial skill, the option of purchasing feedstuffs, fertilizers and other agro-technical inputs, and no limitation on stock water. Irrigation of wheat or pasture is not feasible. Most importantly, the ratio of the price of meat to that of feed is assumed to be above the threshold at which the use of purchased concentrate feeds becomes feasible. This last feature is fundamental to the management of animal production, and may well be a primary classificatory criterion for systems of animal production world-wide. Many of the management options discussed in this chapter only come into existence by virtue of the assumed price regime. This aspect has also been taken into account in the previous chapters where options for greater or lesser emphasis on concentrate feed were defined. The selection of systems was dependent on the ratio of input and output prices that are themselves determined mainly by the terms of trade at the regional boundary.

6.1.3. *Why integrated systems?*

Where grain production is only marginally profitable, and meat prices are not sufficiently high to strongly favour meat production, integrated systems suggest themselves as a way of spreading risks, improving cash flow and gaining a certain degree of synergism. In good rainfall years, the wheat component produces straw which can be utilized by the flock. In poor rainfall years, the flock can graze the wheat when it is green as an alternative to harvesting for grain. This can save supplementary feeding costs. In general, the availability of additional green or dry feed from the wheat component (at virtually no extra cost) means that the stocking rate on the pastoral component can be set at a level that makes better utilization of primary production in the medium and better years. The benefit of integration lies primarily in the grain-producing component acting as a buffer for the meat-producing component. Wheat production integrates well with meat production since it provides a relatively cheap source of nutrients that can be utilized in the target-oriented nutrition of the flock.

6.1.4. *Measures of system performance*

The goal that will be used as a performance criterion in this chapter is maximization of gross margin on the farm, or in other words, the gross margin per unit area, assuming a fixed size of farm. This measure of system performance is of most interest at the farm level of resolution. Various technical efficiencies are also presented since these are of interest at the regional level of resolution. Some of these correspond to efficiency terms used in the pasture system generator (PSG) described in Chapter 5. Notably: total concentrate feed use per unit meat output, total pasture utilization per unit pasture area and per ewe, and meat output per ewe. The relationship between the economic and technical efficiencies will be discussed below.

6.1.5. *Long-term and short-term decisions*

Rainfall unpredictability and variability creates the need to distinguish between long-term and short-term management decisions. The long-term decision is taken independently of the state of the system at the decision time as well as independently of the expected short- to medium-term performance of the system. Such decisions are often related to aspects of system configuration that can not, or only inconveniently or uneconomically, be changed from season to season and can not be determined from the current state of the system or behaviour of driving variables. These decisions in a development context would probably be taken under multiple-goal constraints; they will be discussed in Chapter 7. The short-term tactical decisions, that are the subject of the present chapter, are taken in response to the immediate state of the system and/or in consideration of the expected short- to medium-term performance of the system. In some instances, rational decision making requires a comparison of alternative courses of action where the outcome of each alternative is associated with a relatively low level of uncertainty. In others, the possible outcomes derive directly from weather unpredictability, and an important step in the analysis is the derivation of an outcome-probability function. There are also short-term decisions for which decision criteria can be formulated on the basis of simple mathematical analyses, and it is sufficient to monitor the system so as to have adequate warning of the approach of critical thresholds.

Short-term decisions include:
- the grazing schedule of the ewe;
- grazing deferment;
- early-season grazing of green wheat;
- late-season grazing of green wheat;
- lamb feeding;
- lamb rearing;
- supplementary feeding of the ewe;
- straw baling;
- wheat hay cutting.

In the following subchapters the formulation of the short-term decisions into appropriate optimization algorithms and the integration of these algorithms into an agro-pastoral system model are discussed. The account is non-mathematical, though a few equations have been used for succinctness. A formal definition of the model is given in Ungar (1990). The model can be run for a number of years according to available data (here 21 years). The analysis of these runs provides information on the relative importance of the various management options, the sensitivity of output to management manipulations, and the range of economic and biological efficiencies.

6.2. Short-term decisions

6.2.1. *The grazing schedule of the ewe*

Determining the grazing schedule of the ewe means deciding which of the alternative feed sources should be utilized at any point in time. Six ewe 'locations' are defined in the agro-pastoral system model:
- green pasture;
- dry pasture;
- early-season green wheat (not as an alternative to grain);
- late-season green wheat (as an alternative to grain);
- wheat aftermath;
- holding paddock.

The model decides the location of the ewe at any decision time in two stages. First it determines which of the currently available locations are deemed 'grazable' and then it determines which 'grazable' location to choose.

Determining which locations are 'grazable' involves a series of optimization algorithms which are considered below. Green pasture is 'grazable' from the time pasture biomass exceeds the optimum deferment biomass. Similarly, an optimum stock entry time can be defined for early-season green wheat, which determines when this location becomes 'grazable'. Late-season green wheat is deemed 'grazable' only if it is economically preferable to graze the wheat rather than leave it for grain. Dry pasture is 'grazable' if there is some minimum quantity of biomass in the field, and if the biomass exceeds that of the wheat aftermath. Similarly, wheat aftermath is 'grazable' if there is some minimum quantity of biomass in the field, which also exceeds that of the dry pasture.

Choosing between 'grazable' locations is based upon a user-determined ranking of the locations. The priority ranking used in the standard run in this study is (highest to lowest): late-season green wheat (as an alternative to grain), green pasture, early-season green wheat (not as an alternative to grain), dry pasture, wheat aftermath, holding paddock.

6.2.2. Grazing deferment

Grazing deferment is one of the most important management controls over grazing system dynamics, particularly at higher stocking rates. At even the most abstract level of description it is difficult to discuss appropriate or optimum stocking rates without considering grazing deferment. This management decision is important for three reasons (Chapter 2):
- the net growth rate of a grazed sward is the balance between growth and consumption processes;
- under a given set of environmental conditions, both these processes are strongly related to the quantity of herbage present;
- the balance between these two processes is negative or very small over a wide range of herbage availability and stocking levels.

Grazing deferment is essential if this balance is negative during the initial growth phase. It may also be employed when the balance is positive but small, in order to increase the rate at which availability increases.

The optimum time to commence pasture grazing can be estimated using a simple low-resolution algorithm. It seems reasonable to assume that the stock entry day that maximizes gross margin of the system will be similar, if not identical, to that which maximizes cumulative herbage consumption. Consumption can be defined in terms of green herbage consumption (GC) and dry herbage consumption (DC), weighted according to their relative nutritive value. In the integrated agro-pastoral system, the lower requirement for dry pasture herbage due to the consumption of wheat aftermath (WC) should be taken into account. On this basis an objective function of maximizing consumption (GC + DC + WC) was developed (Ungar, 1990).

The grazing deferment algorithm locates the optimum length of deferment by examining the cumulative green-season herbage consumption for all possible lengths of deferment from zero to the assumed length of the green season (120

138

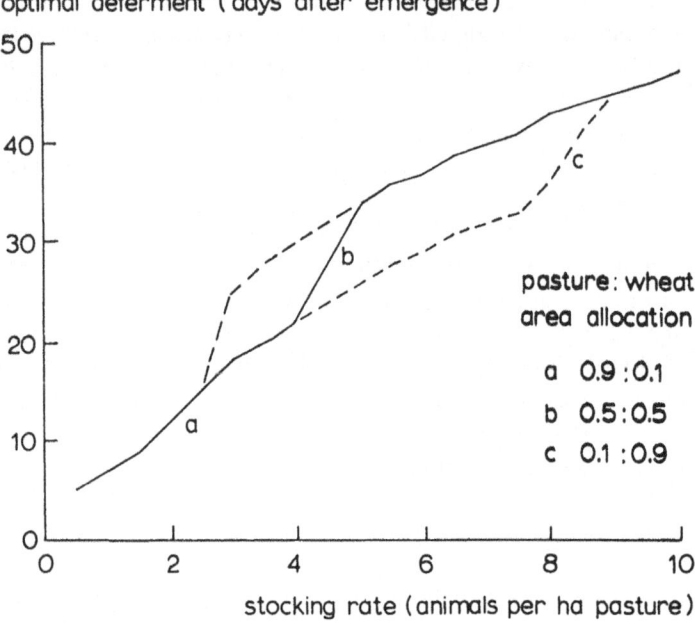

optimal deferment (days after emergence)

pasture: wheat
area allocation

a 0.9 : 0.1
b 0.5 : 0.5
c 0.1 : 0.9

stocking rate (animals per ha pasture)

Fig. 6-1. The optimum deferment of stock entry (days) as a function of stocking rate (animals per ha of pasture), and land allocation between pasture and wheat. Curves a, b and c refer to the situation where, respectively, 90, 50 and 10 per cent of the land is allotted to pasture.

days). A logistic function describes the rate of herbage growth. It was para-metrized from the simulated undisturbed growth curves over 21 years (Section 3.4.2). For each length of deferment, the biomass remaining at day 120 is taken as the dry pasture available at the start of the dry season. The quantity of wheat aftermath expected to be available for the dry season can be estimated from the peak undisturbed biomass assumed in the logistic function for pasture production, since total primary production for pasture and wheat are similar.

Results of the algorithm can be summarized in the relationship between optimum deferment of grazing on pasture and stocking rate, shown in Fig. 6–1. Stocking rate is shown here in terms of the number of animals per unit area of pasture. The optimum deferment refers to the pasture area, and results in the maximum total consumption of green pasture, dry pasture and wheat aftermath. The full drawn curve presents the optimum deferment period for the situation where half of the land is allocated to pasture and half to wheat. At low stocking rates there is a surplus of dry herbage available for use in the dry season, so that the deferment period can be short. However, at high stocking rates the period of deferment has to be long to enable sufficient growth of herbage for later use. The dotted curves show that the transition from short to long deferment shifts to higher stocking rates as the proportion of land that is allotted to wheat increases.

In the agro-pastoral system model, grazing is allowed to commence as soon as pasture biomass exceeds the computed optimum entry biomass or if the number of growing days has exceeded an arbitrary deferment limit (80 days after emergence). It is assumed that if pasture biomass has not reached the optimum by this time then it is probably a disastrous year and there is no point deferring any longer.

6.2.3. *Early-season grazing of green wheat*

Under deferred grazing management, the flock is generally maintained in a holding paddock on supplementary feeds. The cost of feeding can be considerable since this period usually coincides with high pregnancy or early lactation in the ewe. These feed costs can be reduced by grazing green wheat during part of the pasture deferment period. Experiments at Migda indicate that there is a period of at least six weeks from emergence during which defoliation does not reduce grain yield (Benjamin *et al.*, 1976; Yanuka *et al.*, 1981). Beyond this period, defoliation reduces grain yield, the effect on yield increasing with lateness and severity of defoliation (Dann, 1968). Insufficient data are available to accurately quantify the effect of extended grazing on grain yield. In view of this uncertainty it is assumed here that wheat grazed beyond six weeks after emergence is not harvested for grain. Such an option is discussed as a separate decision.

The management decision of early-season grazing of green wheat is whether or not to graze the wheat and when to commence grazing. Experiments at Migda have shown that early-season defoliation reduces peak vegetative biomass by up to five times the quantity of biomass consumed. If the consequent reduction in the availability of wheat aftermath needs to be replaced with purchased feeds, the benefit from early-season grazing may be cancelled. This question will be addressed using the system model.

If the effect on the availability of wheat aftermath is ignored, then the optimum time to commence wheat grazing can be estimated using an algorithm of low resolution. As reasoned earlier, the stock entry day that maximizes gross margin of the system is assumed to be similar, if not identical, to that which maximizes cumulative herbage consumption. Cumulative herbage consumption can be calculated using a simple model based on two functions that define intake and herbage growth rates.

Growth during the first six weeks after emergence can be assumed to be exponential and intake rate is expressed as a function of wheat biomass. The optimum entry day is found by calculating the cumulative consumption till 42 days after emergence for all possible entry days from the decision time. At a given relative growth rate, the optimum entry day increases with stocking rate. At a given stocking rate, the optimum entry day first increases with relative growth rate to a maximum deferment period and then declines with higher values (Ungar, 1990).

6.2.4. *Late-season grazing of green wheat*

An important aspect of sheep-wheat integration is the option to utilize green wheat for grazing as an alternative to grain. The relevant period for this decision commences at the end of the early-season wheat grazing period (approximately six weeks after emergence), and terminates when the wheat crop is ready for harvest. However, in the early phase of the decision period green biomass is probably low, *i.e.* the benefits of grazing are limited, and uncertainty regarding expected grain yield is high. In mid-season, both herbage biomass and quality are relatively high and the expected grain yield can be estimated with less uncertainty. It is during this period that the decision becomes most relevant.

In order to choose between grazing and grain it is necessary to estimate the expected grain yield. In this first analysis, elements of risk are ignored and, therefore, it is only the mean expected grain yield that needs to be estimated. As in the other short-term management decisions, the problem of maximizing gross margin is reformulated in terms that enable the relevant sub-system to be identified and treated using a simple decision algorithm.

6.2.4.1. *Calculating the mean expected grain yield*

The mean expected grain yield is calculated by a form of possible outcome analysis. The possible outcomes are the grain yields resulting from possible future rainfall patterns. Thus the calculation involves generating possible rainfall patterns from the decision time to the end of the season, and the estimation of grain yield from a rainfall pattern. The simplest way of generating possible rainfall patterns is to merge the actual rainfall pattern since the start of the season with historical data for the remainder of the season. For the Migda site, over 20 possible rainfall patterns can be constructed in this way. This series of rainfall patterns can be converted to a set of possible yield outcomes by the use of regression equations or dynamic models.

In this study a regression equation of grain yield on 30-day rainfall is used. The equation is based on rainfall data and wheat yields for the Migda site. Any method of calculating grain yield from rainfall data, including a complex simulation model, could be substituted here. This is not essential to the basic approach.

6.2.4.2. *Choosing between grazing and grain*

The choice between grazing and grain only arises if the combination of current pasture availability and current nutritional requirement of the ewe necessitates the provision of supplementary feed. In order to retain the option of harvesting some grain if conditions improve later in the season, the grazing option is taken with respect to an area of green wheat that would provide ewe requirements for one decision time interval. Thus the wheat is strip-grazed. The decision is re-evaluated at each decision time interval. It is assumed that the option with the lowest net cost is consistent with overall gross margin maximization.

The net cost of choosing grain over grazing is equal to the supplementation cost at pasture (Cp), and the net cost of choosing grazing over grain is equal to the forfeited grain income from an area of wheat that would provide ewe requirements by strip grazing (R_w). A fraction of the wheat area is grazed during the next time interval if Cp>R_w.

The choice between grazing or harvesting wheat for grain depends upon the ratio of expected grain yield to vegetative biomass and not on the expected grain yield alone. Grazing is most likely when there has been good early-season vegetative growth followed by severe moisture stress at a phenological stage that is critical to the determination of grain yield.

6.2.5. *Lamb feeding*

6.2.5.1. *Introduction*
The management decision regarding supplementary and complete ration feeding of the lamb consists of whether or not to provide feed and at what rate. The choice of feed is not considered here; it is assumed that a concentrate feed of high energy-density and protein content is available. Since only the maintenance and liveweight change functions are involved in the growing lamb, lamb feeding can be optimized.

The approach to optimization depends upon whether the system is time-based or product-based. In time-based systems, annual profit is maximized by maximizing the rate of profit generation. This requires identifying the input level at which marginal income equals marginal cost. In product-based systems, there is an income ceiling that cannot be exceeded. Fat-lamb production systems, that produce the lamb 'resource' from breeding stock within the system, fall into this category. The number of lambs sold cannot exceed the number born, and the saleweight also has an upper limit that the market will accept. Hence, annual profit is maximized by maximizing profit per unit output rather than per unit time, and the optimum feeding level is that which minimizes the cost per unit liveweight gain (Q). The fact that time itself may represent a cost in terms of interest and risk does not alter the underlying approach. Such factors can be incorporated into the computation of Q.

6.2.5.2. *Model formulation*
The functional form used to develop the model is that given by ARC (1980), relating scaled energy retention to scaled energy intake (scaling is in multiples of maintenance requirements):

$$R = B (1 - e^{kI}) - 1 \tag{1}$$

where
R is scaled energy retention (-)
I is scaled energy intake (-)
B, k are parameters, defined as functions of diet metabolizability.

142

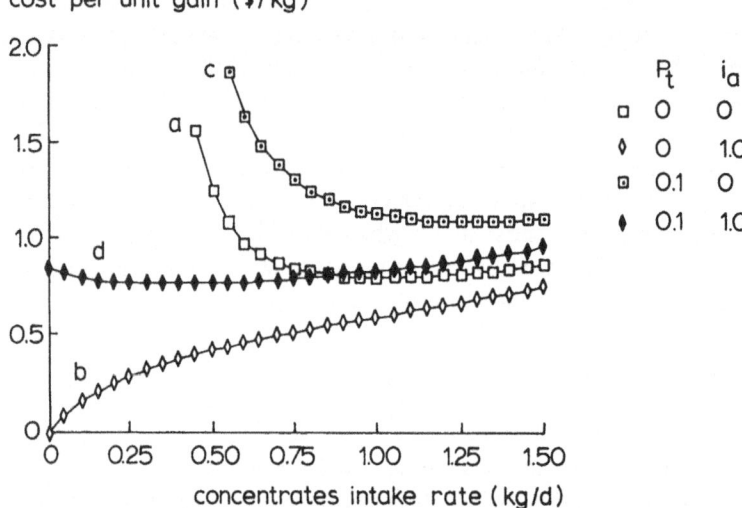

Fig. 6-2. The cost per unit liveweight gain of the lamb as a function of supplementation rate. i_a = herbage intake rate in the absence of supplementation (kg d^{-1}), P_t = time-based costs ($ d^{-1}).

In the first analysis, only feed costs on a single feed diet are considered. In the agro-pastoral system, this would correspond to concentrate-based fattening in a fattening unit. Q is defined as the ratio of feed cost to liveweight gain. It has been shown (Ungar, 1990) that Q is at a minimum when I satisfies the following equality:

$$1-1/B = e^{-kI} (1 + k I).$$ (2)

This expression has to be solved numerically in order to find the optimum I (I*). This solution for a single feed diet can be shown to be identical to maximum biological gross efficiency (see Blaxter and Boyne, 1978).

Outside of a concentrate-based fattening unit, the lamb grazes some form of pasture and may be sucking milk as well. Minimum Q is no longer synonymous with maximum biological efficiency since different feeds with different prices are involved. The computation of Q is more complicated since parameters B and k, and the price of feed energy, change with the diet composition (*i.e.* level of supplementary feeding). A substitution effect, whereby supplementary feed intake displaces pasture intake to some extent, should also be taken into consideration. Here too, the optimum value of I* is found numerically, using a simple algorithm.

A number of time-based non-feed costs are incurred in the process of lamb production and these should be included in the analysis. These costs might include labour, interest, overheads and a risk factor. For the purposes of this analysis, these costs can be lumped together as the time-based costs, converted to a cost per unit gain by dividing by the rate of gain, and incorporated into equation 2 by adding a term to I (for details see eqs. 21 and 22 in Ungar, 1990).

The optimum supplementation level is independent of the price of meat. If it is profitable to continue lamb rearing at all (price of meat > Q), supplementation should be at the level as defined by the above procedure.

6.2.5.3. *Behaviour of the model*

The relationship between Q and supplementation intake rate is shown in Fig. 6–2 for various combinations of pasture intake (in the absence of supplementation) and time-based costs. Consider first the curves relating to a sole concentrate diet. As the supplementation rate increases beyond the maintenance level (at which Q tends to infinity), Q rapidly declines, levelling off as it approaches the minimum, and increases only slowly for supplementation above the optimum level (curve a). The management implication is that in situations where *ad libitum* feeding exceeds the optimum supplementation level, it is safer to over-feed than under-feed when faced with uncertainty. Sub-optimum supplementation can result in Q exceeding the price of meat. The inclusion of time-based costs shifts the cost curve upward (curve c) and raises the optimum supplementation level. At pasture, if some minimal growth rate can be supported in the absence of supplementation and time-based costs are low, then no supplements should be provided (curve b). If pasture intake in the absence of supplementation is insufficient to support growth, or if time-based costs are high, the optimum supplementation level tends to be *ad libitum* (curve d).

The response space of optimum supplementation level to a number of parameters relevant to the calculation of Q shows large regions of zero and *ad libitum* supplementation mediated by a fairly narrow zone of intermediary supplementation levels. It is reasonable to assume that, under field conditions, the system will traverse this boundary region fairly rapidly (*e.g.* increasing pasture availability, increasing time-based costs, declining milk yield) and the management problem of lamb supplementation reduces to a choice between two extreme, easily implemented actions.

6.2.6. *Lamb rearing*

The management problem of lamb rearing consists of selecting a rearing pathway that maximizes profit. The rearing pathway is a nutritional time course, where nutrition is determined by the physical location of the lamb in the system, whether or not the lamb is sucking, and the supplementary feeding regime.

In an agro-pastoral system, eight nutritional locations can be defined:
- holding paddock whilst sucking;
- holding paddock after weaning;
- pasture (green or dry) whilst sucking;
- pasture (green or dry) after weaning;
- wheat (green or dry) whilst sucking;
- wheat (green or dry) after weaning;

from		holding paddock		pasture		wheat		medic	fattening unit
		sucking	weaned	sucking	weaned	sucking	weaned	weaned	weaned
holding paddock	sucking	1	1	1	1	1	1	1	1
	weaned	0	1	0	1	0	1	1	1
pasture	sucking	1	0	1	1	1	1	1	1
	weaned	0	0	0	1	0	1	1	1
wheat	sucking	1	0	1	1	1	1	1	1
	weaned	0	0	0	1	0	1	1	1
medic	weaned	0	0	0	0	0	1	1	1
fattening unit	weaned	0	0	0	1	0	0	1	1

Fig. 6-3. The lamb movement matrix. Defines all possible transfers between lamb nutritional locations in an agro-pastoral system. 0-transfer is not permitted; 1-transfer is, in principle, permitted. Lambs can be born into and sold from any location.

– special purpose pasture after weaning;
– fattening unit after weaning.
The fattening unit and holding paddock for weaners are nutritionally equivalent.

In the development of the agro-pastoral system model it was intended to avoid, as far as possible, the *a priori* definition of rearing criteria. Instead, all possible options are defined, and the algorithm selects between them on the basis of a single economic criterion. The rearing options are contained in the lamb movement matrix which defines the possible flow links between each of the rearing locations. The standard configuration is shown in Fig. 6–3.

Selection of the rearing pathway is based on a comparison of all possible management alternatives, as defined by the lamb movement matrix. Thus the first step in the analysis is to predict lamb performance for each possible alternative, and this needs to be carried out at the optimum supplementation level for the given location, using the algorithm described in Section 6.2.5. The second step in the analysis is to compare lamb performance at the various locations by a single economic criterion. Just as the optimum supplementation level at a given nutritional location is that which minimizes the cost per unit liveweight gain (Q), here also, the optimum location is that which provides the lowest Q (for which rate of income accretion is positive). If no location yields Q < price of meat, or if the maximum saleweight of the lamb has been reached, then the lambs are sold. Under this approach, it is not necessary to set criteria for weaning, supplementation, or lamb sale.

6.2.7. *Supplementary feeding of the ewe*

The problem of supplementary feeding of the ewe is to find the economically optimum supplementation level through time. This is problematic because predictions of productive performance from knowledge of feed inputs are presently less accurate than the determination of the feed input for a given level of performance. The former is only straightforward in the non-pregnant, non-lactating animal, *i.e.* where there is only maintenance and liveweight change. (This enables true optimization of lamb feeding.) Once other productive modes are included, the accuracy of prediction is more restricted.

The difficulty involved in performance prediction is one reason for adopting a target-oriented management approach. Target-oriented feeding is based on input determination since feeding is adjusted to ensure the achievement of specified production targets. These are generally set close to the animal's potential. Thus supplementation policy for ewes is based upon meeting performance targets whenever pregnancy or lactation are involved; that is, outputs for these productive functions are driving variables. Nevertheless, ewe bodyweight is allowed to fluctuate at times during the reproductive cycle when this is not expected to have a detrimental effect on productive performance.

6.2.8. *Straw baling*

The straw baling decision determines the quantity of wheat straw to be baled rather than left in the field. The quantity baled is the biomass that is surplus to the expected dry season grazing requirements. Since baling is soon after grain harvest, the decision is based on expected daily requirements during the dry season. The decision takes into account the quantity of wheat aftermath and dry pasture available, and the 'disappearance' rate of biomass by processes other than grazing.

6.2.9. *Wheat hay cutting*

The option of grazing as a form of late-season utilization of green wheat was discussed earlier. A second alternative to grain is to cut the wheat for hay. Here too, the relevant period for this decision commences at the end of the early-season wheat grazing period, and terminates when the wheat crop is ready for harvest. The decision is based on a comparison of the current value of the standing biomass as hay, and the value of the expected grain yield.

The simplest decision rule is to cut the entire wheat area for hay, if all the following conditions are met:
- the value of the current hay crop exceeds the hay harvesting costs;
- hay is more profitable than grain, assuming the mean expected grain yield;
- conditions do not indicate that hay would be more profitable if cut at the next decision time interval.

6.3. The agro-pastoral system model

The above management decisions were studied in the context of an agro-pastoral system model (Ungar, 1990) which is based on the resources, socio-economic environment and logistic infrastructure of the northern Negev of Israel. It can be used to examine the within-year sensitivity of a farm in this region to management manipulation. In that way it can be used to investigate the behaviour of the systems that are described on an annual input/output basis by the PSG (Chapter 5) and which are selected by the multiple-goal development approach (see Chapter 7).

The model simulates one hectare of land that is divided between pasture, wheat and, optionally, special purpose pasture for lamb fattening. The hectare is exclusive of a holding paddock (which is present in all systems) and a lamb fattening unit (if used). Livestock consists of breeding ewes (including replacer hoggets) and lambs. Rams are not considered. Breeding stock is not bought into the system and stocking rate remains constant between years. Culling time and culling percentage is season-independent and all replacers are drawn from endogenously produced lambs. There is only one breeding season per year. The only feed sources bought into the system are concentrate feed for ewes and lambs, and poultry litter for ewes. Animal nutrition and production are based on an energy balance system. Ewe supplementation is target oriented. Protein requirements are assumed to be satisfied. All prices are in dollars and no inflationary effects are considered. Profit is defined as the gross margin per hectare per annum.

The model is structured as a set of subroutines for the simulation of biological and management functions. The primary production subroutine is based on the simulation model ARID CROP (Section 3.4.2). Adjustments were made to ARID CROP to differentiate between natural pasture, wheat, and medic growth. Computation of animal energy requirements and performance is based upon the feeding system defined in ARC (1980). Reproductive performance is defined by a set of parameters as discussed in Chapter 4. Ewe feeding is target oriented, *i.e.* it is based on providing a nutritional regime that is consistent with a given reproductive performance level. Herbage intake rate is computed from the potential herbage intake rate, the effect of herbage availability and quality (digestibility) on potential intake, and the substitution rate of supplementary feeds for herbage. Potential herbage intake rate is based on energy requirements in the ewe, and the voluntary intake rate of high quality feed in the lamb.

The model is implemented using a daily integration time interval for the biological simulator, and a decision interval of 5 days for the management simulator. Each run of the model is a 21-year simulation using meteorological data for the Migda location for the period 1962/63 to 1982/83. Soil moisture conditions are re-initialized to standard values at the beginning of each season (on October 1). Values for other major state variables provide carry-over effects from one season to the next. Most notably, dry herbage biomass of pasture, wheat and medic, ewe liveweight and body condition, and the quantity of stored hay and straw are not re-initialized between seasons. For this reason, a particular season can yield very different results when simulated singly or as part of a multi-season sequence.

Over 200 runs of the model were executed. The standard run has a land allocation of 50 percent wheat and 50 percent natural pasture, and a stocking rate of 5 ewes per ha system (10 ewes per ha pasture). The number of lambs sold per breeding animal per year is 0.89. The price ratio of meat to concentrate feed for lambs is 10:1.

The standard run corresponds most closely to systems 8 and 9 of the 18 basic systems in the PSG (Table 5–2), on cultivable land with the more intensive concentrate use option. The PSG considers only the pastoral component while the cropping systems are defined separately. They are related to each other in the linear programming model for 'technology selection' and consequently are also related to each other in the interactive multiple-goal development model (Chapter 7). However, in the linear programming models, the interactions between the pastoral and cropping components are not explicitly defined whereas in the simulation model they are.

6.4. Results of the agro-pastoral system model

6.4.1. *The standard run*

Figure 6–4 shows the between-season variability of a number of major performance indices for the standard run. Mean gross margin was 290 $ per ha, with a standard deviation of 150.0 $ per ha. Wheat hay was never cut. Straw was baled in 8 out of 21 seasons; an average of 1,774 kg per ha system (3,548 kg per ha wheat) was baled in these 8 years. Over 70 percent of the straw baled was utilized. This does not necessarily mean that the straw supply exceeded requirement, since the degree of straw utilization depends strongly on the sequence of high and low rainfall years. Average weaning age was 128 days. Average annual herbage consumption by the ewe-lamb unit was 377 kg excluding utilization of baled straw, and 474 kg including straw.

6.4.2. *Wheat grazing and utilization*

In the standard run the ewes grazed early-season green wheat for an average of 24 days per season. The average daily herbage consumption of the ewe during these periods was approximately 0.5 kg per day. The system exhibits 'compensatory' or 'buffering' properties that reduce overall sensitivity to this management option. Thus the very simple decision algorithm for early-season wheat grazing would appear adequate. Under Migda conditions, and at the stocking rate examined, the early-season grazing of green wheat is probably an unnecessary management complication.

148

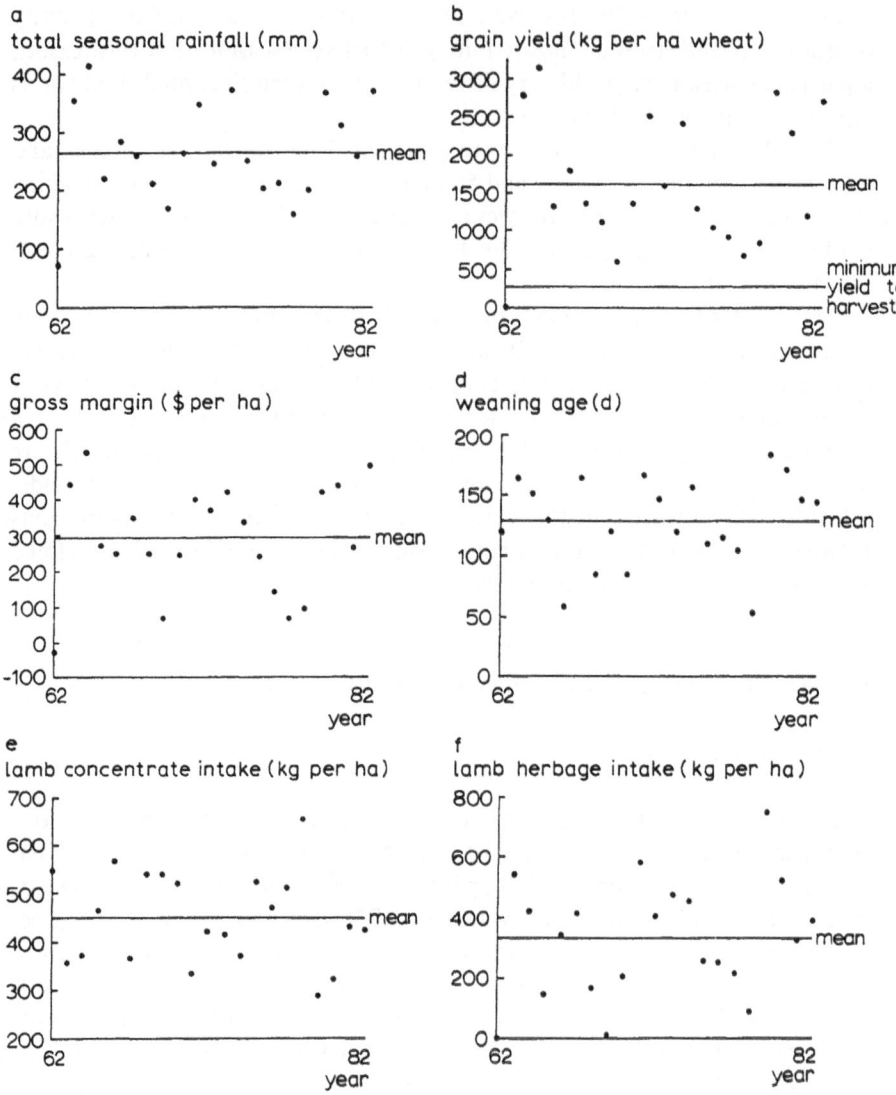

Fig. 6-4. Results of the standard run of the agro-pastoral system model. (a) Total rainfall (mm) (b) Grain yield (kg ha⁻¹ wheat) (c) Gross margin ($ ha⁻¹) (d) Weaning age (d) (e) Total lamb concentrate supplementary feed intake (kg ha⁻¹ system) (f) Total lamb grazed herbage intake (kg ha⁻¹ system) (g) Total ewe grazed herbage plus baled straw intake (kg ha⁻¹ system) (h) Total ewe concentrate supplementary feed intake (kg ha⁻¹ system) (i) Concentrates fed per unit meat sold (kg kg⁻¹) (j) Straw baled per unit area wheat (kg ha⁻¹) (k) Total ewe and lamb pasture intake per unit area pasture (kg ha⁻¹)

An area of green wheat was grazed as an alternative to harvesting for grain in 8 out of 21 years in the standard run. The wheat grazing period lasted for up to 15 days. The largest fraction of the wheat area that was grazed in any one season was 17 percent (in 1977). Wheat grazing did not occur in all years that

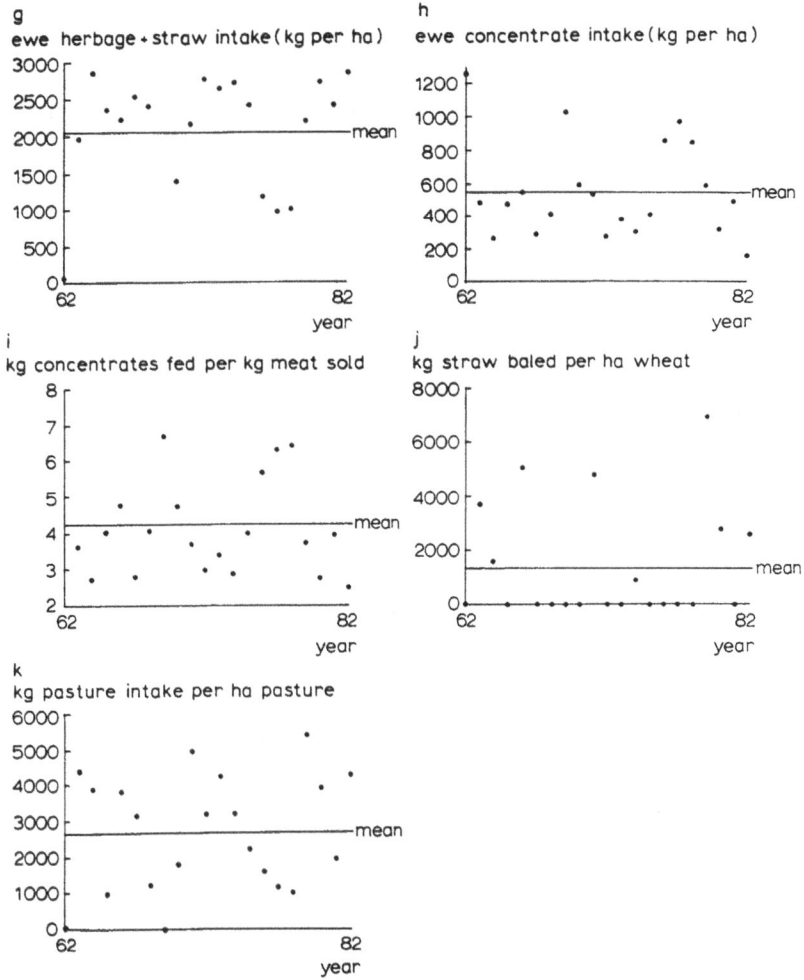

yielded insufficient grain to cover production plus harvesting costs, nor was wheat grazing confined exclusively to years with low grain yield.

Blocking the wheat grazing option reduced mean gross margin over 21 years by only 3 percent. This rather small overall effect on gross margin came about via quite major changes in management pathway in the 8 seasons that were directly affected. On the whole, blocking the wheat grazing option resulted in a delay in weaning, with the lamb receiving a greater portion of its requirements from herbage and less from concentrate feed. However, for the ewes, later weaning in these seasons increased total supplementation and reduced herbage utilization (presumably because of increased lamb herbage utilization). Under Migda conditions, and at a stocking rate of 5 ewes per ha system, the late-season utilization of green wheat by grazing appears to be an unimportant management option.

The wheat area was never cut for hay in the standard run. However, there were instances where the value of the hay crop at the decision time was only slightly less than the mean expected grain profit. Frequent hay cutting does improve technical efficiencies related to concentrate use and pasture utilization, but overall economic efficiency declines due to the loss in grain income and the reduction in the quantity of straw baled.

In the standard run, straw was baled in 8 out of 21 years. On average, the ewes grazed the wheat aftermath for 107 days per season, with a utilization of 433 kg per ha system. The lambs spent an average of 11 days at wheat aftermath, but this was in order to delay weaning, and thereby maintain a relatively low cost per unit gain, rather than for the nutritional value of the herbage.

The three management options of early-season grazing of green wheat, grazing of wheat aftermath, and straw baling are closely related and can be combined in a variety of ways. The early-season grazing of green wheat and the utilization of wheat aftermath by grazing can be allowed or blocked. Straw baling can be blocked, allowed in accordance with the decision criteria outlined in Section 6.2.8, or forced whenever the value of straw exceeds the cost of baling. This yields a total of 12 permutations.

The results (not shown) appear to fall into a group that permits straw baling and a group that does not. Within each group the differences in mean gross margin are small. It is in fact surprising that certain runs yielded such similar results. Once again the property of robustness under quite different management configurations emerges clearly. The variable that correlates most obviously with mean gross margin is the average annual supplementation to the ewes. Two main determinants of ewe supplementation are the energy requirements of the ewe and the availability of herbage for grazing or as straw. A critical parameter in determining the energy requirement of the ewe is the activity increment due to grazing. This reaches a maximum when herbage availability and quality do not limit intake, and grazing activity is at a maximum. Thus the lowest energy requirements are achieved when the ewe does not graze but spends the greatest amount of time off pasture in the holding paddock.

Under Migda conditions, the management options of cutting green wheat for hay, early-season grazing of green wheat, and late-season grazing of green wheat appear to be of marginal importance. The main contribution of the wheat component in integrated systems is the availability of wheat aftermath. Utilization of wheat aftermath is more preferable by baling and feeding in the holding paddock than by grazing. This conclusion indicates that integration of sheep and wheat need not necessarily involve complex management complications. It also indicates that the assumptions underlying the technology selection approach (Chapter 7) are not in conflict with the results of the agro-pastoral simulation model.

6.4.3. *Lamb rearing*

As lamb rearing is the central activity in this study, this area of management will be analysed in some detail. The lamb feeding and rearing algorithms are based on the single economic principle of minimizing cost per unit liveweight gain. The model is not constrained by other criteria in selecting a lamb rearing pathway, and any one of numerous permutations enabled by the lamb movement matrix could, in principle, be selected. A second important feature of the algorithm is that it is based upon the expected costs per unit liveweight gain at any location *at the decision time*. The pathway by which the lamb reached its current position, and the future expected behaviour of the system, are not considered at all in the decision-making process. Despite the simplicity of the decision criteria, the model generally selected conventional rearing pathways.

In the standard run, lambing is on 26 December. This is almost always after the first effective rains and germination. In most cases the ewe is in the holding paddock at lambing time, though in a few seasons the ewes are at early-season grazing of green wheat. Lambing never occurred at green pasture but was always during the pasture deferment period. On the basis of the 21 rearing pathways generated in the standard run, a number of basic patterns can be identified.

In one type of lamb rearing pattern, the lambs suck milk at pasture for the duration of the green season, are weaned at 35 to 40 kg liveweight, and are finished to 45 kg in the fattening unit. Such a pattern is associated with seasons of average or above-average rainfall, with good distribution for sustained primary production once the green season has commenced. This type of rearing pathway was followed in 9 out of 21 seasons.

A second type of lamb rearing pathway occurs in extreme drought years or years with exceptionally poor rainfall distribution. The lambs are, to a large extent, reared in the holding paddock, receiving milk and *ad libitum* concentrate feed. Since herbage availability is low, the lamb rearing options are the holding paddock on *ad libitum* concentrates plus milk, or the fattening unit on the same concentrate diet. The fact that part of the ewe's supplementary feed diet in the holding paddock is used to produce milk is taken into account in computing the costs per unit liveweight gain. This will tend to counter-balance any reduction in costs per unit liveweight gain arising from a higher growth rate on a milk plus concentrate diet than on a concentrate only diet. In each of the four years characterized by this rearing pattern, lamb growth rates were extremely high, and lactation continued till the lambs reached the saleweight of 45 kg.

A third type of lamb rearing pattern can be characterized as sucking at pasture until green herbage availability limits intake, followed by weaning at 25 to 30 kg liveweight, and finishing in the fattening unit. Such a pattern is associated with seasons of below average rainfall. This rearing pathway was followed in 5 out of 21 seasons. In each case, pasture availability became limiting during the green season, and the ewes had to be supplemented at pasture. It is seasons of this third type that pose the most difficult management decisions. In good rainfall years or in serious drought years the rational decision is either obvious or there are few alternative courses of action. In the intermediate seasons, very different management pathways can be triggered by a single decision at a sensitive phase in the season. Here again, system robustness to alternative rational management pathways tends to minimize the financial risk of decision-making under conditions of uncertainty.

It is clear from the above discussion that cost per unit liveweight gain is significantly lower when the lamb is receiving milk. One might, therefore, expect forced early weaning to have a negative effect on overall profitability. However, forcing weaning at 34 days of age reduced mean gross margin by only 7 percent and at 64 days had virtually no effect on mean gross margin. This remarkable degree of robustness was obtained despite large effects on the ewe and lamb management pathway. As the forced weaning age increased, the ewe increased total green pasture consumption, decreased total dry pasture consumption, and increased total wheat aftermath consumption. There was also an increase in time spent in the holding paddock and a reduction in straw baling, due to increased herbage utilization by both ewe and lamb. For the lamb, earlier weaning increased the time spent in the fattening unit and hence total lamb supplementation, but this additional cost was balanced by a reduction in total ewe supplementation.

6.4.4. *Weaning lambs on sown legume pasture*

Although the inclusion of sown legume in the agro-pastoral system is a long-term decision, it is appropriate to discuss this management option together with short-term decisions related to lamb rearing. Viewed in isolation, sown legume swards do possess a number of advantages over non-leguminous swards. However, at the system level, the fact that the area of at least one other component has to be reduced in order to include the legume is in itself a disadvantage. If an area of wheat is displaced by the introduction of legume, then there is a reduction in grain income and straw availability. If an area of pasture is displaced by the legume, then ewe grazing pressure at pasture is increased. This will affect ewe supplementation requirements via the effect on total herbage production and the length of the pasture grazing deferment period. These negative effects would have to be more than compensated by the saving in lamb supplementation achieved by the introduction of a legume area.

Results of the model are given in Table 6–1. Relative to the standard run, an area allocation of 0.45, 0.45, 0.1 to natural pasture, wheat, and medic, respectively, decreased mean gross margin by about 1 percent (R 30). Increasing the medic area to 0.2 and 0.3 of system area, with the remainder divided equally between pasture and wheat, reduced mean gross margin by 3 and 9 percent respectively (R 31 & R 32). Not only did total ewe supplementation increase with increasing medic area, as expected, but total lamb supplementation was

Table 6-1.
Summary of results for the standard run and runs related to the inclusion of medic area in the system.

run number	1	30	31	32	70	71
fraction system area to pasture	0.50	0.45	0.40	0.35	0.50	0.50
fraction system area to wheat	0.50	0.45	0.40	0.35	0.50	0.50
fraction system area to medic	0	0.10	0.20	0.30	0.10	0.20
mean GM $ ha^{-1} system	289.6	286.1	280.2	263.8	292.8	280.2
number of seasons straw baled (-)	8	8	8	6	8	8
total quantity of straw baled kg ha^{-1} system	14,190	11,598	9,519	7,355	13,544	11,131
total quantity of straw utilized kg ha^{-1} system	10,224	8,986	8,284	6,197	10,602	8,194
mean ewe concentr. suppl. kg ha^{-1} system	547	560	573	610	530	595
mean lamb concentr. suppl. kg ha^{-1} system	449	462	468	489	485	512
mean concentr. used per unit meat sold kg kg^{-1}	4.26	4.37	4.45	4.70	4.34	4.73
mean grain harvested kg ha^{-1} system	787	719	635	554	798	800
average weaning age days	128	102	92	80	92	75
mean lamb sale age days	167	173	187	195	174	192
mean time spent at medic days	0	37	63	77	44	76
mean time spent in fattening unit days	58	30	24	27	30	25
mean lamb medic intake kg ha^{-1} system	0	145	259	335	168	301
mean daily medic intake kg day^{-1}	0	0.74	0.78	0.82	0.72	0.75
mean herbage intake by lambs kg ha^{-1} system	330	394	469	497	379	454
mean herbage + straw intake by ewes kg ha^{-1} system	2,041	1,891	1,782	1,586	1,930	1,675
mean lamb + ewe pasture utilization kg ewe^{-1}	268	241	221	194	221	174

also increased in the 3-component systems. This is a surprising result, especially in view of the fact that total lamb herbage utilization and total medic utilization increased with increasing medic area, and the average time spent in the fattening unit decreased with increasing medic area. However, a very significant portion of medic utilization simply replaced herbage utilization at other grazing locations. Furthermore, inclusion of medic resulted in earlier weaning, and so a further portion of medic utilization can be regarded as replacing forfeited milk intake. It should also be pointed out that whilst the average time spent by the lambs on *ad libitum* concentrate feeds in the holding paddock was markedly reduced by the inclusion of a medic area, the lambs did receive supplementation for a significant portion of the time spent at the medic location. This was due to low medic availability or quality late in the season, supporting only very low lamb growth rates and yielding a higher cost per unit gain without supplements than with intermediate or *ad libitum* supplementation.

In R 70 and R 71 (Table 6–1) the wheat area was maintained at 0.5 of total system area, and medic was introduced at the expense of natural pasture area only, but also here increasing the medic allocation to 0.2 of system area reduced mean gross margin by 3 percent. In R 70 the lamb performance figures are quite similar to those in R 30. Once again, total lamb supplementation was increased relative to R 1. However, there was a small increase in straw utilization by the ewes and a small reduction in total ewe supplementation. This tended to cancel the increase in lamb supplementation and thus there was little overall effect on gross margin.

The inclusion of a medic sward in the agro-pastoral system does not markedly improve profitability. Unless there are factors that favour introduction of a medic sward that are not considered in the model, this can probably be regarded as a fairly marginal management option. Some of the other factors that were not considered include the contribution of legumes to the available soil nitrogen and its utilization by the wheat crop. Probably of greater importance is the effect of legumes in a wheat rotation on the phytosanitary aspects of the soil. As a rule, continuous cultivation of wheat cannot be practiced for more than a few years without loss of yield or expensive phytosanitary measures. The nutritive value of legumes and its effect on lamb growth may have been under-estimated in the present model. However, all these factors would have to be quite considerable to change the general conclusion stated above.

6.5. Economic efficiency and technical efficiency

By way of summarizing the results of the agro-pastoral system model, the effects of three major long-term decisions on the overall economic and technical efficiency of the agro-pastoral system were examined. The long-term decisions are stocking rate, land allocation and breed (prolificacy). Stocking rate and breed find reasonably close corresponding terms in the PSG, and are fundamental to system definition there too. Land allocation only finds expression when the results of the PSG are fed into the multiple-goal linear programming package (Chapter 7). Hence it is important to examine the effect of this farm-level management decision on the technical efficiency of the system. In addition, both the PSG and the multiple-goal optimizer ignore the biological inter-annual variation and base all coefficients on mean long-term values. The extent to which these mean values represent a system that is based

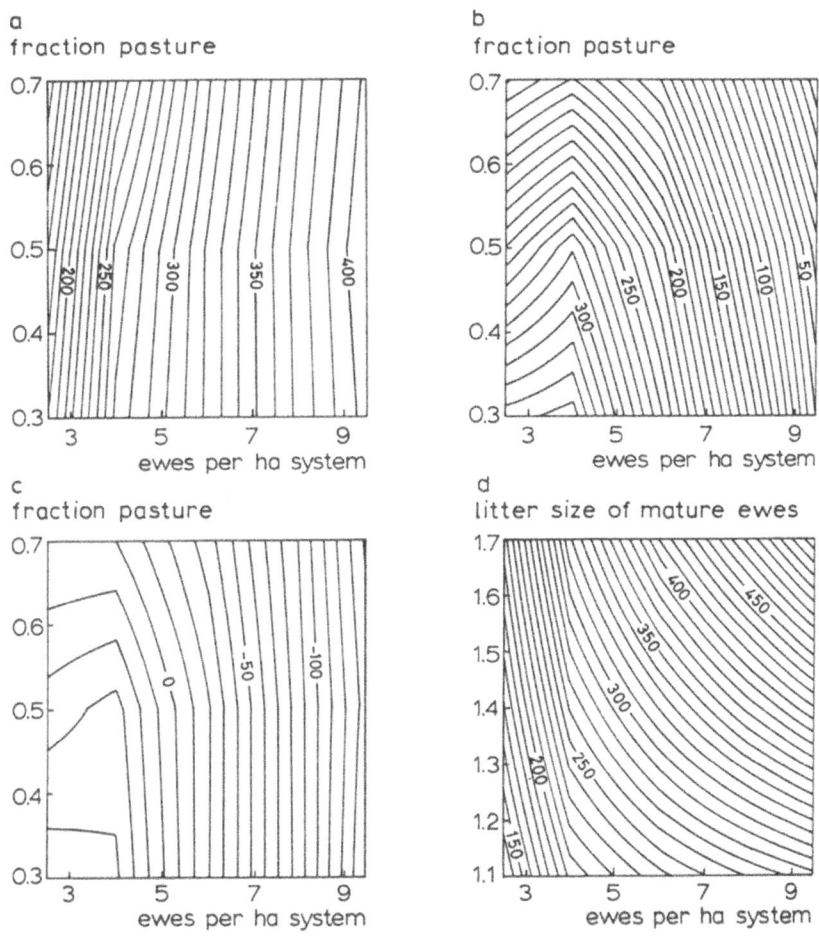

Fig. 6-5. The response surface of gross margin ($ ha⁻¹ system) to long-term management decisions. Gross margin as a function of stocking rate (ewes ha⁻¹ system) and area allocation to pasture (fraction of the system) at a meat:feed price ratio of (a) 10:1 (price of lamb = 2.5 $ kg⁻¹), (b) 5:1 (price of lamb = 2.5 $ kg⁻¹), (c) 5:1 (price of lamb = 1.25 $ kg⁻¹). (d) Gross margin as a function of stocking rate (ewes ha⁻¹ system) and prolificacy (litter size of mature ewes) at a meat:feed price ratio of 10:1 (price of lamb = 2.5 $ kg⁻¹).

on current decisions that change with fluctuating weather and growing conditions can be estimated with the agro-pastoral system model.

The first series of runs was carried out for a meat:feed price ratio of 10:1, with a price of lamb meat of 2.5 $ per kg. The model was run over 21 years for each combination of 5 stocking rates (2, 4, 6, 8, 10 ewes per ha system) and 3 land allocations (0.25:0.75, 0.50:0.50, 0.75:0.25 fraction pasture:fraction wheat) to yield a series of 15 runs. The annual means from these runs were used to construct a response surface (Fig. 6–5) of various efficiency measures to stocking rate and land allocation.

156

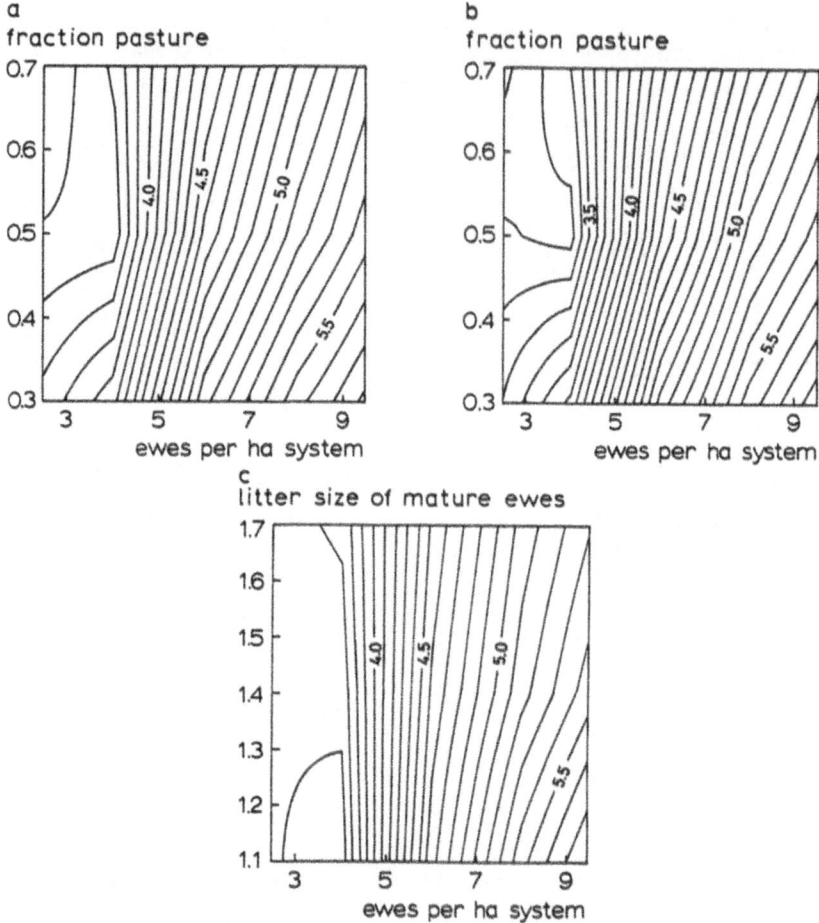

Fig. 6-6. The response surface of efficiency of concentrate feed use (kg concentrate fed to ewes and lambs per kg ewe and lamb meat sold) to (a) stocking rate (ewes ha⁻¹ system) and area allocation to pasture (fraction of the system) at a meat:feed price ratio of 10:1 (price of lamb = 2.5 $ kg⁻¹), (b) stocking rate (ewes ha⁻¹ system) and area allocation to pasture (fraction of the system) at a meat:feed price ratio of 5:1 (price of lamb = 2.5 $ kg⁻¹), (c) stocking rate (ewes ha⁻¹ system) and prolificacy (litter size of mature ewes) at a meat:feed price ratio of 10:1 (price of lamb = 2.5 $ kg⁻¹).

The second and third series of runs were carried out as above for a meat:feed price ratio of 5:1. In the second series the price of lamb (liveweight) was $ 2.5 per kg and the price of feed $ 0.5 par kg; in the third series $ 1.25 and $ 0.25 per kg respectively.

In the fourth series of runs the response surface is for stocking rate and prolificacy. The same five stocking rates were used as before in combination with 3 levels of fertility (1.0, 1.4 [standard run], 1.8 lambs per lambing for mature ewes; the litter size of hoggets was adjusted in proportion). Prices were as in the standard run (and the first series).

a
fraction pasture

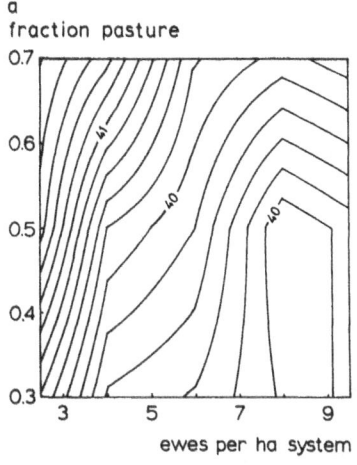

b
litter size of mature ewes

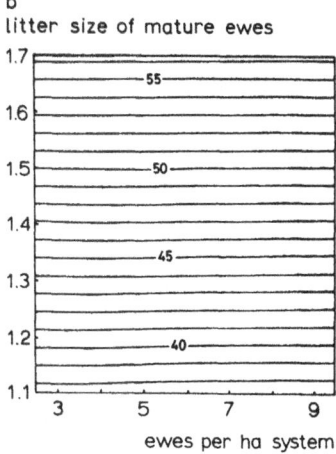

ewes per ha system   ewes per ha system

Fig. 6-7. The response surface of the quantity of ewe and lamb meat sold per ewe (kg) to (a) stocking rate (ewes ha⁻¹ system) and area allocation to pasture (fraction of the system) at a meat:feed price ratio of 5:1 (price of lamb = 1.25 $ kg⁻¹), (b) stocking rate (ewes ha⁻¹ system) and prolificacy (litter size of mature ewes) at a meat:feed price ratio of 10:1 (price of lamb = 2.5 $ kg⁻¹)

Figure 6–5 shows the gross margin response surface for the four series. In series 1, gross margin is quite insensitive to land allocation at all stocking rates. In series 2 and 3, sensitivity to land allocation is low at high stocking rates only. The optimum stocking rate is about 4 ewes per ha irrespective of land allocation. In series 4, gross margin increases with stocking rate and litter size over the entire range examined. Gross margin climbs rapidly with stocking rate up to 4 ewes per ha and less steeply thereafter. However, sensitivity to litter size is least at low stocking rates.

These very different response surfaces of economic efficiency are in sharp contrast to those obtained for technical efficiency. Figure 6–6 shows the efficiency of meat production in terms of concentrate feed use for series 1, 2 and 4. Results for series 2 and 3 are identical. Results for series 1 and 2 differ slightly at low stocking rates. Whilst economic efficiency increased (higher gross margin) with increasing meat:feed price ratio, this index of technical efficiency is lower (higher concentrate use per kg meat produced) at the higher price ratio. This is because lambs are sold slightly lighter at the lower price ratios. The entire range in efficiency of feed use for all four series is approximately 3 to 6 kg feed per kg meat sold. Sensitivity to land allocation is generally very low.

The response surfaces for series 1, 2 and 3 are identical for pasture utilization per unit area pasture and pasture utilization per ewe, but the response surfaces for series 1 and 4 were different. Land allocation has a greater effect on these measures of technical efficiency than does prolificacy. Pasture utilization per unit area pasture is most sensitive to land allocation at high stocking rates, but even then the variation is only 10 to 15 percent. Thus the fact that possible effects of integration with wheat are ignored in the PSG is not a serious problem. However, the absolute level of utilization assumed in the PSG may be a little high. This over-estimation may result from ignoring between-season variability in pasture productivity and utilization. As variability increases, mean utilization will decrease. This is because utilization will decrease when production is less than potential utilization in poor years, but will not increase when production is greater than potential utilization in good years.

Meat production per ewe at a price ratio of 10:1 is constant with stocking rate and land allocation since the cost per unit gain never exceeds the price of meat before the saleweight is reached. Thus no response surface of meat output per ewe is shown for series 1. The response surfaces for series 2 and 3 are virtually identical since it is the price ratio and not the absolute prices that determines whether or not to continue lamb rearing. The response surfaces for series 3 and 4 are shown in Figure 6–7. Both stocking rate and land allocation had some effect on meat output per ewe at the low price ratio, but the total variation over the entire response surface is very small. The response to litter size and stocking rate at the high price ratio is as one would expect under target-oriented management.

In conclusion, it has been shown that large changes in economic scenarios that materially change gross margin from negative values to over 400 $ per ha can have only a small effect on most of the important technical efficiency criteria of agro-pastoral systems. This is a characteristic of fairly complex systems with strong negative feedbacks between components and lends validity to the averaging approach adopted in the long-term multiple-goal analysis.

7. Modelling agricultural development strategy

I. SPHARIM, R. SPHARIM and C.T. DE WIT

7.1. Introduction

Development within an agricultural region can be defined as a change in infrastructure and technology that is undertaken to improve the productivity and welfare of the community. A 'technology' is a well defined production activity, whereby inputs are converted into products. There are many technologies available, but not all are feasible in the physical and socio-economic environment of a given region. The 'socio-economic environment' in the present context is expressed by the regional constraints and prices of inputs and products.

Analysis of options for regional agricultural development in a dynamic, integrated manner can be done with a suitable multiperiod, multiple-goal linear programming model (Hadley, 1982). Such a model selects mixes of technologies that will satisfy the development objectives subject to the regional constraints. The model simulates technological development in different regions and the effect of different socio-economic scenarios on technology selection and scheduling within a region. In this way, robust characteristics of a development path, that is, aspects that are not sensitive to change in scenario, can also be identified. Finally, the model is able to test the potential impact of technologies based on innovative techniques that are still being tested under experimental conditions. Selection of these technologies could then serve as a guide for selecting research proposals.

The multiperiod characteristic of the model simulates the time dimension of development and traces feasible development paths. This can provide regional planners with the capability for scheduling activities at the most appropriate time during the development process. The interactive multiple-goal mode provides a means whereby tradeoffs between different goals can be evaluated.

Any agricultural development model is by definition a simplification of a complex socio-economic, physical and biological system. Such a model aims to capture only those elements of the system that are relevant to the purpose of the study. Here the main purpose is to select those technologies that accord with the prevailing physical and socio-economic environment and best promote the development goals of the region and of other legitimate stakeholders. Such a model should have at least the ability:

Th. Alberda et al. (eds.), Food from Dry Lands, 159–192.
© 1992 *Kluwer Academic Publishers.*

160

a. to select a set of products and production techniques that are best suited to the requirements of the region.
b. to represent the effect of external conditions (mainly reflected by the prices of products that are exported from the region and costs and availability of inputs that are imported into the region) on the selection of activities within the region.
c. to allow the necessary interaction between activities within the region.
d. to simulate the generation of financial capital as well as the investment, obsolescence and replacement of physical capital during the course of development.
e. to account for the effect of availability and reproduction potential of genetic stock on technology selection.
f. to identify technically feasible pathways for medium to long-term development processes.

All of these capabilities can be built into a linear programming model based on activities, constraints and goals. Activities represent production technologies that convert resources (inputs) into products or services (outputs). The resources include natural physical and biological resources that exist in the region and also inputs that are obtained from outside the region. Products or services can be used within a region or traded across the regional boundaries. The activities not only draw on limited resources (or constraints), but can also contribute to them. For instance, income generated in the region can be used to increase capital.

Activities can transfer quantities over time from one period to the next or in the same period from one production system to another. The first mode of transfer is characterized by activities such as capital formation. For example: financial capital in one year is transformed into physical capital in the next year, and this contributes capital services in the following years. A different type of capital formation is related to livestock population dynamics: female lambs born in one year are transformed into ewes that increase the livestock capital in the following year. The ewes produce products (lambs) or capital (female breeding lambs) and so forth. The other mode of transfer is of products produced by one activity as an input to another activity. For example, the grain producing activities, which can transfer the products grain and straw to the sheep husbandry activities where they are used as feed inputs.

Trade activities are channels that allow trade across the region borders. These include purchases of imported inputs, including fertilizers, concentrates or hired labour, as well as sales of products that are exported out of the region. Cash flow activities are those that relate to management and utilization of money for investment or consumption by the population in the region. The money for investment can be produced within the region, or borrowed from a lending agency or bank from outside the region or obtained as a grant from an international aid agency.

Constraints can also be divided into different categories. There are constraints which are natural resources that are not changed endogenously in the course of development. An example is total land area. This does not preclude the possibility of land reclamation that can change land productivity. Other constraints are resources that change as a result of development within the region. Examples are physical capital that can be increased by diverting part of the income generated to investment; or breeding stock that can increase by reproduction. A third category of constraints include those, like concentrate feed (fodder grains) or fertilizer, that can be produced within the region but can also be imported from outside. In the first and second constraint categories, the total available resources can be fully or partially exploited by the selected activities. In the third category of constraints, inputs are bought as needed and so the allocation among all participating activities is complete and the total for the region is balanced. For example, fertilizers are purchased according to the requirements of the selected activities, and there is no excess or deficiency of fertilizers, subject to the availability of money to purchase them. Another example is the constraint related to revenue that is allocated either to investment or to consumption and does not remain unused.

The goals, or targets, of development are defined either as outputs of activities that are to be maximized (like income) or constraints that can either be maximized or minimized. These include employment (maximize), fertilizer use (minimize), income for consumption by the people in the region (maximize), and so on. The treatment of multiple goals that may be imposed on a region will be discussed in greater detail later in this chapter (see Subchapters 7.3 and 7.4).

7.2. The multiperiod regional development model

The technique of linear programming (LP) requires that a set of relevant 'technologies', similar to those defined in Chapter 5, is translated into an equivalent set of 'activities'. Each activity is a set of coefficients (a vector) that relates all the relevant inputs of a specific production process to its outputs. The same conceptual pattern is used to define trade and capital formation activities which convert products to money via sale or purchase activities and revenue from sales to physical capital (buildings and equipment).

The activity vector describes the input/output relations per unit of a relevant production factor. Thus, cropping activities are normalized to the unit of land; livestock activities to the appropriate animal unit; investment, borrowing and consumption activities to the unit of money. The values of the target functions and the values of the solutions are also related to the unit to which each activity is normalized.

The linear programming method calculates the mix of activities that can use the available resources (constraints) most effectively to meet the goal. This goal or target function, can be the total revenue for consumption by the people in a region over a development period; or the number of settlers in a region subject to a minimum income constraint; or the need to reduce dependence on foreign aid; or the requirement to minimize the use of a particular polluting fertilizer.

An iterative procedure of multiple-goal optimization that does not assume that relative weights for each goal are known in advance will be considered in Subchapter 7.3.

The 'mix of activities' refers to the 'amounts' of each of the activities (or technologies) that are selected to maximize the target without violating any of the resource constraints. If such a mix can be formulated, then there is a 'feasible' solution. If not, then there is no feasible solution, unless some of the constraints can be made less severe. Expressed in concise mathematical forms, the LP problem is solved by maximizing (or minimizing) a target function, Z, subject to a set of constraints:

$$\text{Max } Z = x' * c$$

subject to $Ax \leq b$
$$x \geq 0$$

where,

x is the solution vector that contains the amount of each activity. The amount of an activity cannot be a negative number, and, therefore, each value in x is greater than or equal to zero.

c is a vector that defines the contribution of each activity to the selected target function.

A   is the activity matrix, a matrix of technical coefficients with m constraints (rows) and n activities (columns), with n > m. The technical coefficients define the input or resource requirement per unit of product.

b is the vector of constraints, commonly called the right hand side (RHS). This defines the available resources and as a rule requires that they set an upper (or lower) limit to the utilization of that resource.

One of the main requirements of a development program is the scheduling of activities over a development path, while taking into account investment opportunities, obsolescence of physical capital and reinvestment policy. For this purpose the activity matrix of the linear program is elaborated into a multiperiod grand tableau. This grand tableau defines the interrelations between the production, trade and capital formation activities in the course of time. In the example of Figure 7–1, a development horizon of 15 years is considered. The diagonal elements A (1,1) – A (15,15) are the activity matrices for each year, whereas the non-diagonal elements A (2,1), A (3,1) – A (3,2) up to A (15,1) – A (15,14) handle the transfers of animals, physical and financial capital from one year to the next. Any physical assets, like fences, are assumed to be obsolete in a period of 7 years, so that there are at most 7 elements in a row or column of the grand tableau (Fig. 7–1). By means of this multiperiod configuration a mix of activities for each year is generated that optimizes the objective function for the whole time-horizon under consideration. In this way an optimum course of development is obtained in which some production activities fade out, others move in and still others stay the same or are not used at all. It thus identifies dynamically feasible and scheduled development paths.

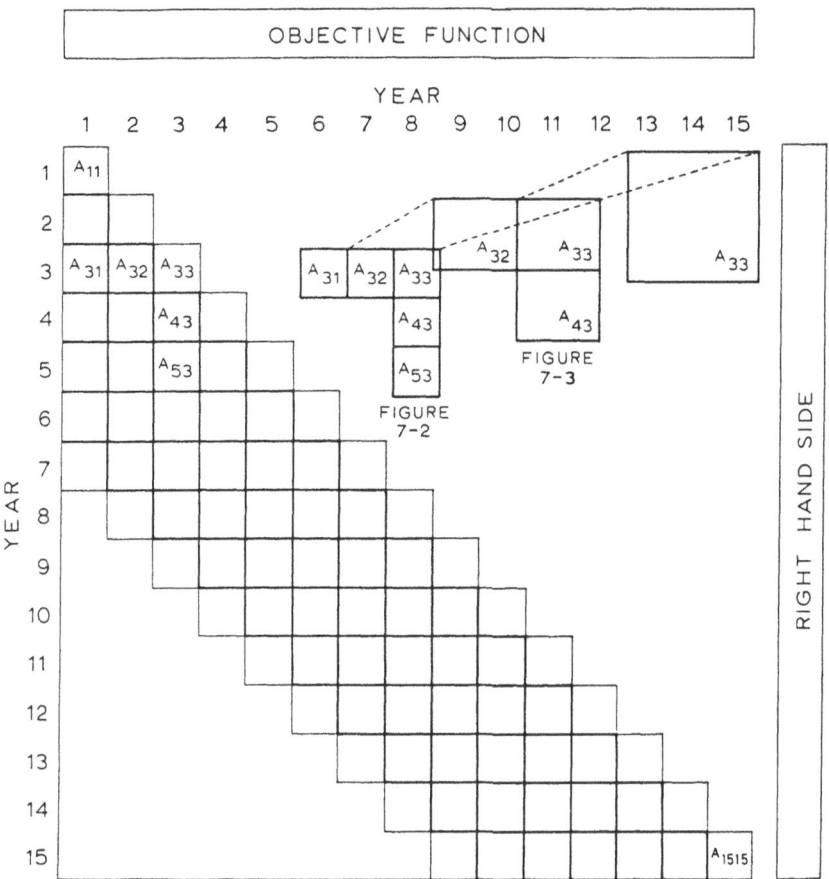

Fig. 7-1. Layout of the multi-period linear programming 'Grand Tableau'.

A closer view of the tableau for three consecutive years is given in Figure 7–2. The activities (columns) are divided into six main sections (I – VI) and are referred to by capital letters A-M in Figure 7–3; the constraints (rows) are divided into 13 categories and are referred to by the numbers 1–13. The activity matrix for one of the years and its links to adjacent years are presented in Figure 7–3. Activities are given here for the years (t-1) and (t), and constraints for the years (t) and (t + 1). The submatrix that is formed by the columns and rows for the year (t) is the activity matrix for that year, whereas the other two submatrices concern the transfer of animals and physical and financial capital from one year to the next. The detailed structure of the basic activity matrix, the 'A-matrix' is discussed in detail in Section 7.2.1.

In Figures 7–1 to 7–3, presentation of the actual coefficients for each activity is impractical in a schematic diagram because of the large number (more than a hundred) of production activities. The coefficients for these activities are calculated with the pasture system generator (PSG) described in Chapter 5 and

		year 1						year 2						year 3					
		Production	Capital formation &/or use	Intermediate products	Hired labour	Cross-border trade	Cash flows	Production	Capital formation &/or use	Intermediate products	Hired labour	Cross-border trade	Cash flows	Production	Capital formation &/or use	Intermediate products	Hired labour	Cross-border trade	Cash flows
		I	II	III	IV	V	VI	I	II	III	IV	V	VI	I	II	III	IV	V	VI
Natural resources	1																		
Real capital	2		-1						-1						-1				
Labour force	3																		
Animal feed	4																		
Fertilizer	5																		
Other inputs	6																		
Animals-ewes	7								-1					+1					
Animals-breeders	8													-	+1				
Mutton	9																		
Main production: grain	10																		
By-product: straw	11																		
Money capital	12														+1				
Revenue	13																		1.05
Natural resources	1																		
Real capital	2														-1				
Labour force	3																		
Animal feed	4																		
Fertilizer	5																		
Other inputs	6																		
Animals-ewes	7														-1				
Animals-breeders	8																		
Mutton	9																		
Main production: grain	10																		
By-product: straw	11																		
Money capital	12																		-1
Revenue	13																		
Natural resources	1																		
Real capital	2														-1				
Labour force	3																		
Animal feed	4																		
Fertilizer	5																		
Other inputs	6																		
Animals-ewes	7																		
Animals-breeders	8																		
Mutton	9																		
Main production: grain	10																		
By-product: straw	11																		
Money capital	12																		
Revenue	13																		

Fig. 7-2. Transfer matrices that enable the program to allocate resources from one year to the next and so facilitate optimization of the development process.

here it is only necessary to describe the relationship between the activities and the constraints. This is done with the following conventions: The symbol (+) in an element indicates that the activity concerned draws upon the constraint concerned and the symbol (−) that the activity contributes to the constraint. The constraints that act as units to which other inputs and outputs are related are indicated by either (+ 1) or (− 1). In the present study there are animal units (ewes) in the pastoral activities, land units (ha) in the cropping activities and money units ($) in the trade and cash flow activities.

SECTION KEY				year t - 1													year t												
I Production; II Capital formation &/or use; III Intermediate products; IV Hired labour; V Cross-border trade; VI Cash flows				I		II		III		IV	V		VI				I		II		III		IV	V		VI			
				Animal systems	Crops	Buildings & equipment	Hoggets for reproduction	Animals	Crop products	Hired labour	Purchases	Sales	Investment	Aid	Borrowing	Consumption	Animal systems	Crops	Buildings & equipment	Hoggets for reproduction	Animals	Crop products	Hired labour	Purchases	Sales	Investment	Aid	Borrowing	Consumption
				hd	ha	$	hd	hd	kg ha	my	kg	kg	$	$	$	$	hd	ha	$	hd	hd	kg ha	my	kg	kg	$	$	$	$
	CONSTRAINTS			A	B	C	D	E	F	G	H	I	J	K	L	M	A	B	C	D	E	F	G	H	I	J	K	L	M
year t	Natural resources	ha; m3	1														+	1+1											
	Real capital	$	2			-1											+	+	-1										
	Labour force	my	3														+	+					-						
	Animal feed	fu	4														+					-		-					
	Fertilizer	kg	5														+	+						-					
	Other inputs	$	6														+	+						-					
	Animals-ewes	hd	7				-1										1+1												
	Animals-breeders	hd	8														-			1+1	1+1								
	Mutton	kg	9														-				-				+				
	Main product: grain	kg	10															-				+			+				
	By-product: straw	ha	11														+	-											
	Money capital	$	12										-1						1+1								-1	-1	
	Revenue	$	13												1.06								+	+	-	1.05			1+1
year t +1	Natural resources	ha; m3	1																										
	Real capital	$	2																-1										
	Labour force	my	3																										
	Animal feed	fu	4																										
	Fertilizer	kg	5																										
	Other inputs	$	6																										
	Animals-ewes	hd	7																	-1									
	Animals-breeders	hd	8																										
	Mutton	kg	9																										
	Main product: grain	kg	10																										
	By-product: straw	ha	11																										
	Money capital	$	12																							-1			
	Revenue	$	13																									1.06	

Fig. 7-3. The basic activity matrix, 'A', and its links to adjacent matrices.

7.2.1. *Activities and constraints*

The activities and constraints of Figure 7–3 are discussed here according to their main categories. The activity categories are the following:

Production activities. These are the actual and potential technologies that are relevant to the development of the region and include the animal production and cropping systems discussed in Chapter 5.

Capital formation activities. These represent the formation of physical and biological capital. Physical capital is formed by drawing on money capital and using it for structures, building and equipment acquisition. Biological capital is the genetic stock used for reproduction. It is formed by drawing on young female animals (potential breeders) in one year and converting them into reproductive breeding animals in the following year.

Intermediate products. These include:
- Animal products. This activity allows for the conversion of young potential breeders into meat products for sale.
- Crop products. This activity allows grain produced by the cropping activities to be sold or used as animal feed.

Hired labour (employment activity). This activity makes it possible to hire labour in addition to the permanent labour force in the agricultural sector of the region. In this way it contributes to the labour resource (or constraint) by drawing on revenue in the region. Because labour supply can be related to employment goals it is defined separately from the following group of activities.

Cross border trade. These are activities that allow import of services and products that serve as inputs to production activities and for export of products outside the region.

Cash flow activities. These include:
- Investment, which is the withdrawal of financial capital (money) from revenue in one year for investment in physical capital in the following year.
- External (foreign) aid that contributes to the available financial capital, *i.e.* the money available for investment. This is an external factor which allows for injection of funds to stimulate development. It does not have to be paid back to the contributor.
- Borrowing. This accounts for loans from financial sources not directly related to the production process in the region. It contributes to investment funds in one year, but must be returned with interest from available revenue in subsequent years.
- Consumption. This activity represents 'consumptive income', which is defined in the present study as income before taxes less investment. It draws on revenue and makes it available to the people in the region for consumption of products and services from outside the region or outside the agricultural sector.

The main constraints are:

Natural resources, like land (ha), that can be divided into categories relevant for the production activities of the region. Another limiting resource could be the amount of drinking water.

Real (physical and biological) capital, that includes fences, buildings and equipment already in the region at the beginning of the development horizon or acquired in the course of development. It also includes capital on the hoof, *i.e.* the ewes and the young breeding females (hoggets) that can be sold for meat or be retained and become breeding ewes in the following year.

Labour force, which refers to the permanently settled inhabitants of the region who conduct the production activities. They are considered as entrepreneurs who do not earn any wages but participate in the consumption of the revenue. As a constraint this labour force can be relaxed by hiring labour at the current wage rate from outside the region or the agricultural sector.

Tradable materials are concentrates, fertilizers and other miscellaneous inputs that are bought according to need, so that their stock is kept at zero.

Final intermediate products include lamb and mutton for sale as meat, main crop products, crop by-products and roughage.

Financial capital (money) is available for investment and can be adjusted by banking activities.

Revenue is divided between consumption, investments or saving. In years with very little revenue it is possible to borrow money to maintain the consumption level.

7.2.2. Animal production activities

The animal production activities are normalized on the ewe (Figure 7–3: ' + 1' in element A7 in year t). They draw (+) on natural resources (pasture land), physical capital (fences, structures, equipment), labour, fertilizer (for pasture fertilization) and can also draw on the locally produced grain or straw. They contribute (-) products for sale ('mutton') or young stock (hoggets) for breeding up the flock (Figure 7–3: '-' in elements A8 and A9 in year t). The crop production activities are normalized on a unit of land (Figure 7–3: ' + 1' in element B1 in year t). They draw on similar constraint categories as the livestock systems (except for 'ewes' and 'animal feed') and contribute to the crop products, wheat grain and straw (Figure 7–3: '-' in elements B10 and B11 in year t). The use of the wheat for sale or as animal feed is determined by the appropriate intermediate and trading activities F and I.

7.2.3. *Interactions with the socio-economic environment outside the region*

The boundaries of the region are determined by a logistic barrier that can be measured by the cost of transporting inputs into the region and products from the region to the market outside. These costs are influenced by institutional constraints (import quotas, duties), distance and the logistic efficiency of the transport system. They affect different products differently, some becoming untradable, others not. Tradable products not only include those that can be sold from the region, but also products that compete with local production. Among the items defined as 'non-tradable' in this model are the work of the permanent settlers, cultivable land and rangeland, and the breeding ewes. The tradable products are covered by the two trading activities: purchases (H) and sales (I) in year t. Purchases draw on revenue (' + ' in element H13 in year t) and contribute to animal feed, fertilizers and other inputs ('-' in elements H4, H5 and H6 in year t). Sales contribute to revenue ('-' in element I13 in year t) and draw on mutton (lambs or hoggets) and grain (' + ' in elements I9 and I10 in year t). Note that the intermediate animal products activity (E) draws on hoggets (' + 1' in element E8 in year t) and contributes to the 'mutton' that is for sale ('-' in element E9 in year t), the 'mutton yield' of each hogget depending on the specific livestock production technique. Male lambs are contributed by the animal production activities directly to the 'mutton' constraint, together with replacement ewes ('-' in element A9 in year t). Grain export is represented by the sales activity drawing on the grain constraint (' + ' in element I10 in year t) and contributing to revenue ('-' in element I13 in year t). The prices involved in these transactions are those current at the 'region gate'. Import of grains (for concentrate used as animal feed) is represented by the purchasing activity drawing on revenue (' + ' in element H13 in year t) and contributing to animal feed ('-' in element H4 in year t). Here, too, prices are those current at the 'region gate'. The difference between import and export prices of tradable products is an index of the 'height' of the logistic barrier.

7.2.4. *Interactions between activities within the region*

In agricultural production it often occurs that by-products are obtained in addition to the main product. Whereas the main product has a relatively high value per unit weight or volume, can cross the logistic barrier and thus be traded across the region boundaries, the by-products are bulky and of relatively low value. They can be used within the region as inputs for other production activities. These by-products can be important in the functioning of a region when they play the role of obligatory inputs for some activities. An example in this case would be straw that is used as supplementary roughage in certain sheep husbandry systems. In such cases the availability of straw can be a critical factor in determining the intensity of these activities. It is, therefore, important to take this interaction between activities into account. This is done as follows: The cropping activity contributes wheat grain and straw (' −' in elements B10 and

B11 in year t). The straw is used by the animal husbandry activities that draw on the straw constraint ('+' in element A11 in year t); the crop product (intermediate) activity can draw on the grain yield ('+' in element F10 in year t) and contribute grain (as concentrate feed) to the animal feed constraint ('−' in element F4 in year t).

7.2.5. *Capital formation*

Agricultural development depends on enlargement of the means of production. This is done by increasing the physical and biological capital. The formation of physical capital includes the construction of structures and buildings and the acquirement of equipment (activity C). It draws on investment capital ('+1' in element C12 in year t) and converts (or contributes) it to physical capital ('−1' in element C2 in year t). Physical capital can be used without extra cost till it becomes obsolete. Thus in any year there are capital assets of various 'vintages' that appear as physical capital constraints. As long as physical capital is not obsolete, it can be used by various production activities, depending on how specific the capital item is. The wider its potential use by a large number of activities, the lower its specificity. This will be reflected by the symbol '+' appearing in a large number of activities that draw on a particular item of physical capital. In the model most sheep husbandry systems can use fences, which are non-specific, but only a few can use high technology artificial rearers.

In agricultural systems where the reproductive rate of the genetic stock (or biological capital) is low (like in livestock or dates), the availability of suitable breeds or cultivars can be a crucial constraint to rapid development. Where the logistic (or veterinary) barrier prevents trading in genetic stock, its availability must be taken into account. The specific feature of the genetic stock is that it can be sold to increase current revenues, or it can be kept to increase the available biological capital and so increase revenues in the following years. The model takes these two aspects into account in the following way: The animal husbandry systems contribute hoggets or young female lambs that can potentially be used for breeding when they become mature in the following year ('−' in element A8 in year t). The biological capital formation activity (D) draws on the store of breeding animals ('+1' in element D8 in year t) and contributes them to the reproductive flock in the following year ('−1' in element D7 in year t + 1). However, the intermediate 'animal product' activity (E) can also draw on the available breeding animals (ewes and hoggets) ('+1' in element E8 in year t) and contribute them to the meat for sale (mutton) constraint ('−' in element E9 in year t). These animals are then not available for breeding in the following year. The sale activity (I) then draws on the mutton constraint ('+' in element E9 in year t) and contributes to revenue ('−' in element I13 in year t).

In this way the competition for hoggets between sale and breeding is maintained over the whole planning horizon, and ensures that whatever development path is selected, it will be feasible in terms of available genetic stock.

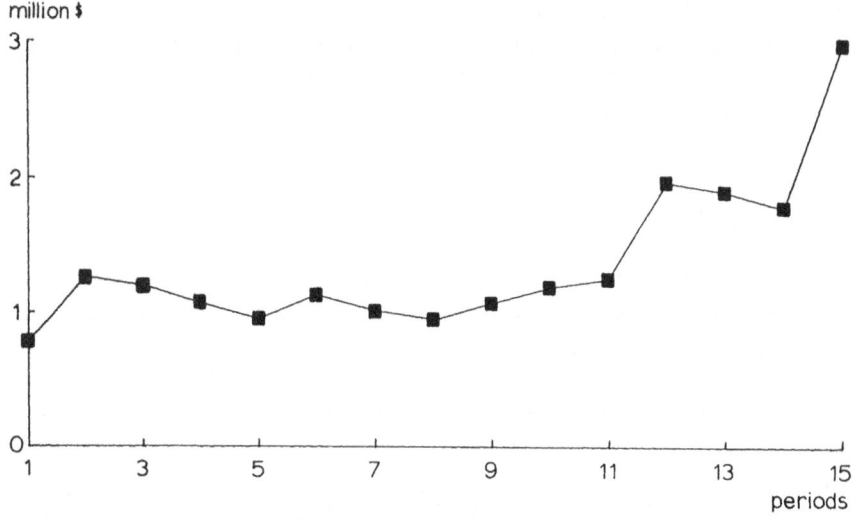

Fig. 7-4. Dynamics of regional consumption of revenue.

7.2.6. *Cash flow budget and time preference*

Capital formation requires investment which is financed from savings within the region or from external sources. The decision whether to invest, to consume or to save the available revenue will depend on many factors, one of the more important of which is the value of time which is determined by the attitude to consumption in the present in relation to investment, with the expectation of greater revenues for consumption in the future. The decision taken is also influenced by the cost of borrowing money and the benefit gained from saving for later investment in the production activities. The availability of money for investment in capital formation and the regional 'cash-flow' over the development horizon will, therefore, depend to a large degree on these factors.

The 'cash-flow' activities (VI, Figure 7–3) include investments necessary for new physical capital (structures, buildings and equipment, (J)), availability of external development aid (K), borrowing for investment through commercial loans (L) and consumption of revenue (M). The investment activity draws on revenue in any one year (' + 1.05' in element J13 in year t) and contributes to investment capital in the following year (' − 1' in element J12 in year t + 1). It is assumed that the population in the region has a time preference for present consumption, so the investment activity bears a cost that draws more (' + 1.05' in year t) than it contributes (' − 1' in year t + 1). The external (or foreign) development aid activity (K) contributes to investment capital (' − 1' in element K12 in year t, the year in which the aid is given). This aid is in the nature of a grant and does not have to be paid back in subsequent years.

The borrowing activity (L) bears a cost of commercial funding from outside the region at the market interest rate, say 6 percent. This activity contributes to investment capital (' − 1' in element L12 in year t) and draws on revenue in the following year (' + 1.06' in element L13 in year t + 1). The consumption activity draws on revenue (' + 1' in element M13 in year t).

7.2.7. *Model output*

A regional scenario was defined and the model was run to gain an impression of the way it operates, and in particular, how some aspects of development change over the planning horizon. This scenario, the 'standard run', represents boundary conditions similar to those in the northern Negev of Israel and the selected target is 'maximum revenue for consumption'. It is, therefore, interesting to examine the course of revenue generation for consumption and investment as well as the regional dynamics of sheep breed composition.

The course of consumption is presented graphically in Figure 7–4. During the first years of development, regional consumption does not increase and even falls in some years as revenue is heavily used for investment instead of consumption. Investment drops to a more modest level as development proceeds. Consumption then rises dramatically towards the end of the planning horizon, whereas investments decline accordingly. If a project is to be liquidated at its termination, then this short term behaviour is a realistic result and opens up opportunities for investment in other projects. However, where regional development is expected to continue, the results that relate to the final years are a 'distortion' and should be ignored. This 'distortion' occurs because towards the end of the planning horizon, it becomes less desirable to make long-term investments and so more revenue is allocated to consumption, (or investment in other projects). In order to map out a longer development pathway, a longer planning horizon can be chosen, the practical period depending only on computer capacity. With a fifteen year planning horizon and continuing development, the model results are applicable for planning purposes only till about the eleventh year of the fifteen.

Selection of sheep breeds suitable for the regional development scenarios is based on the assumption that increase in genetic stock, in this case breeding ewes, can come about from local reproduction and to a limited extent only from imported animals or other more specialized technologies (artificial insemination or embryo transfer). The existing native landrace is the Awassi fat-tail used mainly for both lamb and milk production. As this standard run does not include sheep milk production as a tradable activity, the Awassi is at a disadvantage in comparison to the meat breeds, the German Mutton Merino (GMM) and the Finn cross (with Awassi and GMM). Consequently, the Awassi, which dominates in the first years of development gradually loses its dominance to the GMM and Finn cross, that are available in relatively small numbers at the beginning of development (Fig. 7–5). In the earlier years, the Awassi is fully exploited, while the GMM and the Finn cross are increased to the limit of their biological potential. In the eighth year, the number of GMM ewes exceeds the number of Awassi ewes for the first time, while the Finn cross

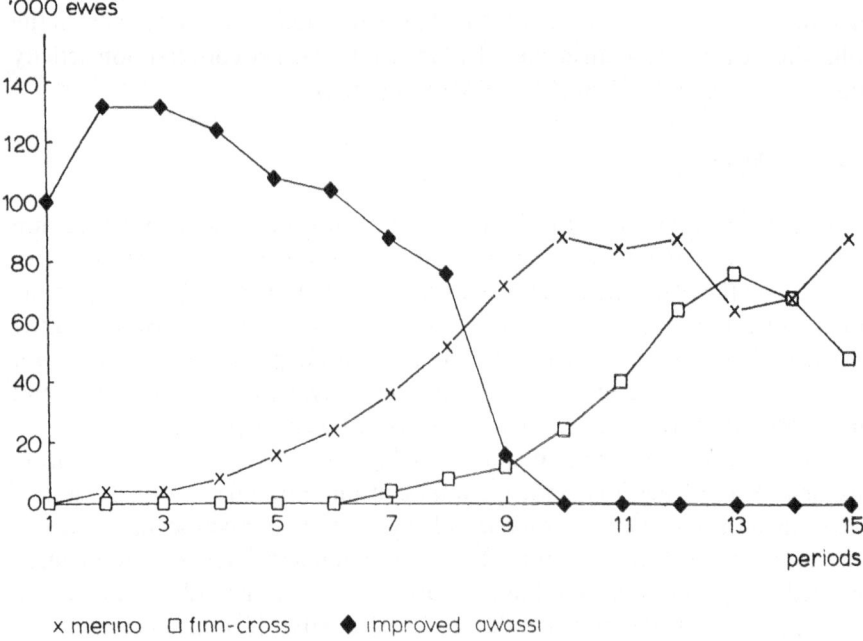

'000 ewes

x merino   □ finn-cross   ◆ improved awassi

Fig. 7-5. Dynamic changes in breed mix over the development horizon.

becomes dominant later on. The flock composition in this run is the result of maximization of regional consumption. It is to be expected that consideration of other goals will change the sheep breed dynamics, but it is of interest to note that in the northern Negev the trend to higher fertility breeds has become evident in recent years.

7.2.8. *Testing model performance*

Can the model respond in a reasonable way to different socio-economic scenarios? In other words, do the results inspire enough confidence to justify their use as guide-lines for regional development and research planning? In order to check whether the model as defined in the A-matrix is operating as intended, some test runs were done to see if it could reproduce realistic technology mixes for widely different socio-economic situations and how it responds to smaller changes within such situations.

7.2.8.1. *Differentiation between regions*
Three different socio-economic regions were defined that roughly reflect the situation in the northern Negev of Israel, a region in the West Australian wheat belt and the coastal strip of the western desert of Egypt. All three areas are semi-arid with a mediterranean type climate. Dryland farming and extensive animal husbandry are the major agricultural practices. The population density is greatest in the Egyptian region and lowest in the Australian region; the distance

Table 7-1.
Definition of three typical regions.

Socio-economic characteristic	'Negev'	'W. Australia'	'W. Desert'
Prices - outputs ($ kg⁻¹)			
Local market price, lamb	2.4	1.5	2.4
Export price of lamb	1.7	1.2	1.7
Export price of wheat	0.16	0.15	0.16
Prices - inputs			
Imported concentrates ($ kg⁻¹)	0.15	0.17	0.15
Phosphorus fertilizer ($ kg P⁻¹)	0.40	0.60	0.40
Nitrogen fertilizer ($ kg N⁻¹)	0.30	0.60	0.30
Hired labour (10³ $ person year⁻¹)	10.0	10.0	5.0
Regional constraints			
Permanent settlers ($ person year⁻¹)	400	100	1,000
Local market for lamb (kg mutton)	1x10⁷	0.5x10⁷	2x10⁷

from markets is greatest in the Australian region and smallest in Israel. The local market for lamb is largest in the Egyptian region. In order to keep the test conditions relatively simple and open to fairly direct analysis, model performance was checked in relation to one goal only: maximization of revenue for consumption. The socio-economic environment is defined in terms of the prices of outputs and bought inputs, and in terms of regional constraints like the number of 'permanent settlers' and the local demand for lamb and mutton. This is a highly simplified definition of a region, but should be sufficient to illustrate the effect of regional characteristics on technology selection for development. A more detailed definition is given in Table 7-1. To simplify the comparison, the region is 50,000 ha in all three cases. The results of the runs are given in Table 7-2.

The model produces strikingly different results and technology mixes in the different socio-economic regions. Total regional income in the 'W. Australian' scenario is lowest, but income *per capita* is the highest. The cropping system there is overwhelmingly a wheat/legume rotation whereas in the 'Negev' and the 'W. Desert' it is mainly continuous grain cropping. In the 'W. Australian' scenario, the use of nitrogen fertilizer is lowest while the use of phosphorus fertilizer is highest; no imported concentrate feed is given at all, while in the other regions, it is used heavily. Whatever concentrate feed is given for fat-lamb production comes from the locally produced grain.

Despite the simplistic nature of this exercise and the fact that alternative feasible technologies like wool or sheep milk production were not included among the available technologies, the results of these runs reflect current development trends in the northern Negev of Israel and in the north-western coastal zone in Egypt. The trends suggested by the model for western Australia are much less realistic, because there the main relevant activities are based on

Table 7–2.
Summary of main results of 'typical region' development features (mid-development phase).

Item	'Negev'	'W. Australia'	'W. Desert'
Cumul. consumption (10^6 $) [1]	244	64	282
Wheat area (10^3 ha yr^{-1}):			
- Continuous wheat	9.12	0	10.91
- Wheat/fallow rotation	0	1.15	0
- Wheat/legume rotation	7.82	17.07	4.82
Total wheat area	16.94	18.21	15.73
Straw area usage (10^3 ha yr^{-1}):	13.03	9.25	13.32
Concentrate imports (10^3 tons)			
- Obligatory concentrate	69.06	0	67.9
- Concentrate, replaceable by legume	2.64	0	6.3
Fertilizer use (10^3 tons):			
- Nitrogen	1.06	0.33	1.24
- Phosphorus	0.53	0.68	0.46
Grain utilization (10^3 tons):			
- Exported from region	20.71	10.05	20.70
- Used in the region	0 7.	84	0
Lamb and mutton marketing (10^3 tons)			
- Exported from region	4.33	5.79	0
- Used in the region	9.92	0.50	14.93

[1] Cumulative consumption over the 15-year development horizon.

fine wool Merino sheep in which wool is the main product and mutton is a by-product. These activities were not considered in the present version of the model. Including them requires data on the Australian sheep husbandry budget, that can then be translated into a vector of technical coefficients. That would enable planners to use the model to explore innovative technologies, including some that these days are being discussed intensively by scientists and pastoralists in Australia. The technologies concern the introduction of dedicated fat lamb production systems based on the Awassi breed that is particularly suitable for the live sheep export trade to the Middle East. This, as well as many other hypothetical alternative production systems could be assessed with the model before investing the huge amount of capital involved. In the same way, this model can be adapted to other situations, such as semi-arid regions in South America or the southern parts of the USA.

7.2.8.2. *Sensitivity to socio-economic scenarios*
The central purpose of the model is to select from a wide choice of technologies, a mix that will best serve the development goals in a specific socio-economic environment. The application of the model in a multiple-goal mode will be discussed in detail in the following subchapter. Here the impact of different regional socio-economic scenarios, specified by changes in the price ratios on the technology mix, will be examined.

Table 7–3a.
Parameter specifications for sensitivity runs (Prices in $ kg⁻¹).

Run	1	2	3	4	5
Wheat for export	0.20	0.20	0.20	0.20	0.20
Imported concentrate	0.17	0.22	0.17	0.17	0.17
N-fertilizer	0.60	0.60	0.60	0.30	0.60
P-fertilizer	0.60	0.60	0.60	0.60	0.40
Lamb local market	2.20	2.20	1.70	2.20	2.20
Lamb export	1.70	1.70	1.70	1.70	1.70

Table 7–3b.
Summary of results of sensitivity runs.

Run number	1	2	3	4	5
Scenario [1]	Stand	Conc (+)	Lamb (-)	N-fert (-)	P-fert (-)
Cumulative consumption (10⁶ $)	205	184	144	209	207
CNO Concentrate use-obligatory (10³ kg)	56.4	47.7	46.1	64.3	56.5
CNR Concentrate use-replaceable (10³ kg)	0.06	0	0	15.8	0.2
FERN Fertilizer use-nitrogen (tons)	623	469	490	828	643
FERP Fertilizer use-phosphorus (tons)	669	717	727	664	674
Wheat exported from region (10³ ton)	22.5	0.9	20.6	25.5	22.9
Wheat used as feed (10³ ton)	9.45	8.68	8.66	9.63	9.46
Wheat area - continuous (10³ ha)	5.98	0.37	3.25	10.82	6.57
Wheat area - wheat/fallow (10³ ha)	0.77	0	0.28	0	0.61
Wheat area - wheat/legume (10³ ha)	13.17	18.62	16.83	10.35	13.01
Total cultivated area (10³ ha)	19.92	18.99	20.36	21.17	20.19
Straw area usage (10³ ha)	13.55	9.68	12.02	16.00	13.87
Meat sold (10³ $)	12.4	8.5	11.4	13.4	32.3

[1] The deviations from the standard scenario are as follows:
Run 2 - higher concentrate price
Run 3 - lower lamb price
Run 4 - lower N-fertilizer price
Run 5 - lower P-fertilizer price
Details in Table 7-3a

As in the previous example the model was run to maximize revenue for consumption. The price changes are given in Table 7-3a and the solution to each of these changes in Table 7-3b. There are five scenarios. The baseline scenario is based on a physical and socio-economic environment appropriate for the northern Negev of Israel. Two scenarios (runs 2 and 3) represent more difficult economic conditions because of increased concentrate prices or decreased lamb prices while the ratio between the unit cost of concentrate and

176

the unit price of lamb (1:10) is held constant. The last two scenarios represent better economic conditions for the producer because of cheaper N or P fertilizer.

The results (Table 7–3b) show that in the socio-economic environment of the northern Negev, lower lamb prices or higher concentrate feed prices tend to increase the desirability of the wheat/legume rotation, reduce the use of concentrate feed and of nitrogen fertilizer, and increase the use of phosphorus fertilizer. However, even though the lamb production profitability is lower, less wheat is produced because of the greater area under legume pasture. It is also of interest to note that even though the unit price ratio between imported concentrate feed and lamb was maintained constant, the resultant cumulative consumption and the selected technology mix were different. In particular, the area of wheat and consequently the quantity of wheat exported, was much lower when the price of imported concentrate was increased than when the price of lamb was reduced.

When the price of lamb and the price of imported concentrate both go down so that the same ratio is maintained, it does not mean that the profitability of production remains intact, because the price of locally produced inputs remains unchanged. This has two effects which in economic jargon are termed the 'expansion effect' and the 'substitution effect'. The lower profitability of lamb reduces lamb production and expands wheat production. The change in price ratio between imported inputs and locally produced inputs increases the substitution of imported concentrates by locally produced legumes.

Some other results are less surprising. Thus, for instance, the lower price for nitrogen fertilizer results in less of the cultivated land being used for the wheat/legume rotation and more for continuous wheat (Run 4). The reduction in the price of phosphorus fertilizer has virtually no effect on the solution (Run 5). This is a consequence of having defined activities with only one fixed level of phosphorus utilization, thus limiting the capacity of the model to respond. The effect of a price reduction in this case is, therefore, mainly to increase the income available for regional consumption.

The economic environment can change during a development project, generally within predictable bounds. Under such conditions of limited uncertainty, it is helpful to know how sensitive the desired development path is to such changes. The scenarios listed in Table 7–3a are distinguished from one another by changes in one factor at a time, and do not truly represent the more complex changes in scenario that involve a number of factors simultaneously. Nevertheless, they do indicate which characteristics are robust and which are sensitive to change. Among the robust characteristics is the development of breed composition:

– In all the runs, the Awassi gives way to the more prolific breeds, and would be maintained only if there were other non-economic reasons to do so.
– The area sown to wheat is not very sensitive to change.
– Only a very small area, if at all, is sown to the wheat/fallow rotation.

These examples show that under conditions of limited uncertainty, many elements of regional development are fairly robust and need not necessarily require major revision with moderate change in the economic environment.

7.3. Interactive multiple-goal planning

7.3.1. *Background*

In most situations where it is necessary to discriminate between different development possibilities there are several objectives and goals that must be satisfied, but since it is usually impossible to realize them all, they must be compromised, either explicitly or by default. According to Veeneklaas (1990), predictive planning strongly relies on insight into the underlying relationships between objectives and on an accurate assessment of existing restrictions, but in many planning situations these requirements cannot be met. Optimization and conventional planning require *a priori* weighting of objectives or permitted deviations from targets. When objectives or targets are expressed in different units of measurement and when, moreover, the objective function must be linear, the weighting procedures become too rigid and easily lead to artificial and consequently unsatisfactory constructions. Compromise planning does not assume that targets or relative weights for each objective are known in advance. The emphasis is on the generation of feasible and efficient technical solutions and active involvement of the policy maker in the decision process. The main task of the technician is to reduce the number of alternatives to a manageable level, while keeping a low profile in the process.

A so called 'satisfizing' (satisf[ying-optim]izing) method of reducing the number of alternatives is described by Simon (1955). A policy maker supplies minimum values he is willing to accept for each objective, and instead of seeking a best solution, he only seeks for a set of good solutions. Alternatives that do not satisfy his minimum requirements are not considered any further. Those that do, are again screened after increasing the minimum acceptable values of one or more of the objectives. When used in an iterative fashion, the number of alternatives can be reduced to very few and, if desired, to a single choice.

Such ideas were taken up by Hartog *et al.* (1979) in their study of the limits of the Welfare State and from there on further elaborated into an Interactive Multiple-Goal Programming (IMGP) procedure by Spronk and Veeneklaas (1983), van Driel *et al.*, (1982, 1983) and Veeneklaas (1990) to support the policy-oriented studies of the future, carried out by the Netherlands Scientific Council for Government Policy.

7.3.2. *A graphical illustration of Interactive Multiple-Goal Programming (IMGP)*

In the IMGP procedure the policy maker is presented with a number of iterations. Each iteration consists of a series of optimizations, the first iteration requiring at least as many as there are goal variables. These may typically be between five and ten. However, following van Driel *et al.* (1983), the procedure can be illustrated with a simple example using only two goal variables, consumption and employment, in the context of a region that fully depends on agro-pastoral activities. It is assumed that the policy maker wants to maximize both goals, even though they may conflict with each other. Therefore, in the first iteration consumption is maximized, but without setting any lower limit to employment and employment is maximized, but without setting any lower limit to consumption. The result is given in Figure 7–6a in the units 10^4 \$ and person-years. Since the values in this example do not have any significance on their own, they are further quoted without these dimensions. Point A presents the maximum consumption of 1,000. The employment happens then to be only 300.

Point B represents the maximum employment of 1,000 but then consumption happens to be only 100. If points A and B coincide, both goals are completely tied, so that realization of one brings with it the realization of the other. There is then no conflict of interest.

Point P1 (300,100) represents consumption when employment is maximized and employment when consumption is maximized. These two values are chosen as the lower bounds of the two goal variables, because it can be guaranteed to the policy maker that he does not have to accept lower values for these goals. The point I1(1,000, 1,000) in the figure combines the highest possible consumption with the highest employment and may be considered the ideal combination. It is, however, an utopian solution, since it is impossible to realize the maximum of two independently formulated, partially conflicting, goals at the same time. The rectangle P1AI1B contains, therefore, more than all possible alternatives for production. To illustrate this, the points A and I1 and B and I1 are joined by dotted lines.

All alternatives to the left of and below the lines P1A and P1B can be replaced by better ones. For instance, by moving from point P' to the right one arrives at an alternative which is always preferred because employment is higher at the same consumption. Therefore, by increasing the minimum goal restrictions to the point P1 no solutions that could be acceptable for the policy maker are excluded.

Given the ideal but utopian combination I1 and the most unfavourable combination P1, the policy maker is asked now which of the lower goal restrictions he wants to increase and to what extent. It should be pointed out to him that he does not commit himself because any value can be reconsidered. Let us assume that he first wants to ensure that consumption is at least 750. The most unfavourable combination of goal achievement is then P2 (300, 750) as in

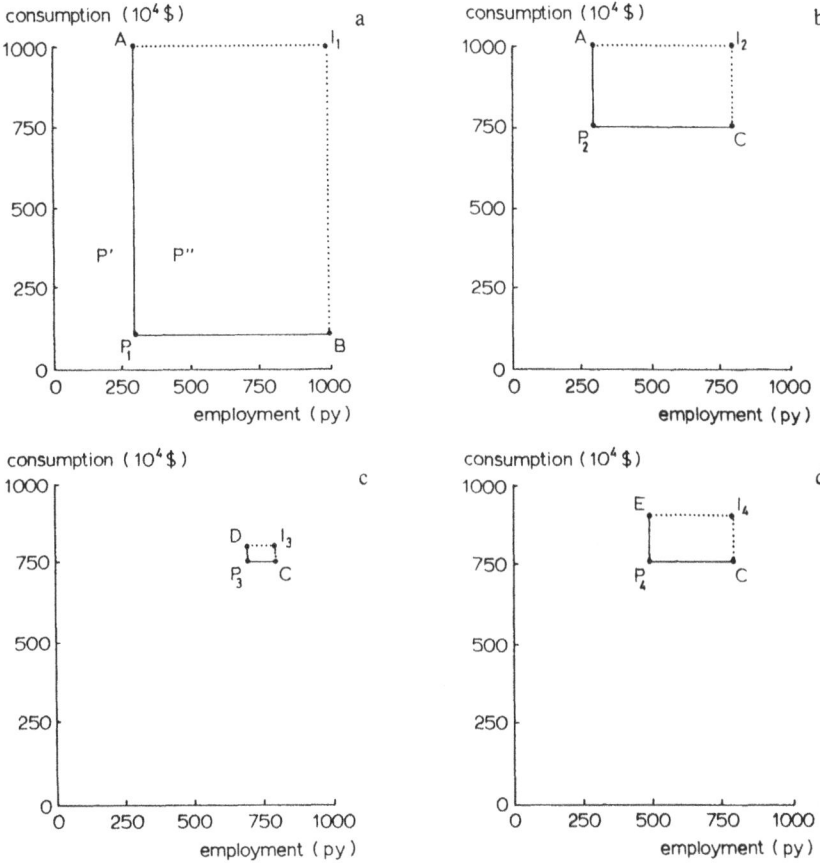

Fig. 7-6. a. specific production possibilities in case of no goal restrictions;
   b. specific production possibilities under the goal restrictions that the consumption is at least 750 and the employment at least 300;
   c. specific production possibilities under the goal restrictions that the consumption is at least 750 and the employment at least 700;
   d. specific production possibilities under the goal restrictions that the consumption is at least 750 and the employment at least 500.

Figure 7-6b. This tightening of the goal restriction has its price and to determine this, it is necessary to execute a second iteration. The constraint that consumption has to be at least 750 does not change the possibility to reach the value of 1,000 at point A. Therefore, that optimization does not have to be repeated. However, it may appear now that the maximum employment that can be achieved is only 800 so that the original combination of goal values represented by point B in Figure 7-6a has to be replaced by point C (800, 750). The ideal or utopian alternative then moves to I2 (800, 1,000). This is the price that has to be paid for increasing the lower limit for consumption. If the policy maker considers this decrease in maximum employment acceptable, the tightened goal restriction for consumption is accepted, and he is asked again to tighten one of his lower limits.

Let it be assumed now that he increases the minimum acceptable employment to 700. Because this is lower than 800, this is a feasible demand. The most unfavourable combination of goal achievements now becomes P3 (700, 750) as in Figure 7–6c. It may then appear that the maximum consumption that can be realised is only 800, as given by point D in this figure. The ideal or utopian point moves then from I2 (800, 1,000) in Figure 7–6b down to I3 (800, 800) in Figure 7–6c. The feasible alternatives are now restricted to an employment between 700 and 800 and a consumption between 750 and 800.

It may be that the policy maker considers the reduction in possible consumption from 1,000 to 800 too large a sacrifice after all to guarantee that employment does not fall below 750. If so, he may reduce the lower limit for employment to for instance 500, *i.e.* halfway between the 300 that was considered too low and the 700 that resulted in an unacceptably low consumption. The points P4 (500, 750) and I4 (800, 900) in Figure 7–6d characterize his new margins.

The policy maker may continue until he arrives at a solution where the points P and I coincide and that he may consider satisfactory. However, he may also want a more detailed analysis of the future situation under those conditions. For that purpose he may request information on the animal husbandry techniques selected when consumption (point E in Figure 7–6d) or employment is maximized (point C). If he is not satisfied with this technology selection, he may want to start again at some preceding step, but with additional goal restrictions, such as the number of animals to be kept in feedlots. To improve his insight into the dependencies, he may also want to examine the shadow prices, that indicate how much the maximum of one goal decreases if the lower limit on another goal is increased by some small amount.

At this point, the investigator may want to re-examine the technical coefficients of the production techniques selected to verify that he has not been overly optimistic with respect to their performance, and if necessary to improve on them. In this way the interactive exercise that guides the policy maker in his regional planning also guides the investigator in his research planning.

The procedure is in principle the same if more than two goals are considered, but then the number of optimizations needed to arrive at a satisfactory solution increases rapidly with the number of goals. With three goals, already six iterations are necessary to tighten each of the goal restrictions once and per iteration the number of optimizations is equal to the number of goals. Since dynamic programs are in general time-consuming, available computing time may become a serious constraint when the activity matrix is large and the planning horizon is long.

It could be attempted to tighten two goal constraints at the same time, but then there may be no feasible solution for the optimization problem. In our example that would have been the case if, for instance, the minimum consumption would have been set at 810 and the minimum employment at 760 at the same time (see Figure 7–6c). The problem with such unfeasibilities is that they are expensive, because it takes the solution algorithm often considerable effort to find this out.

Another possibility to reduce the number of iterations considerably is not to reset a goal after its optimization but to retain a more satisfactory value. In that way the feasibility of subsequent optimizations in the iteration round is ensured, but the danger exists that the policy maker is guided so rapidly to a solution corner, that all options which may be of interest are not explored. Experience shows that this danger is limited if two full iteration rounds have been completed, but by then the most time consuming and expensive optimizations have been performed.

7.4. An example of the application

7.4.1. *The initial conditions*

The method was applied to the northern Negev region of Israel, a mediterranean region, with winter rainfall as explained in Section 7.2.8. To define the region for the present purpose, both cultivable land with a high production potential, usable for both arable farming and rangeland and uncultivable hilly rangeland, were assumed to occupy 25,000 ha each. The initial number of settlers in the region was, rather arbitrarily, set at 400. The genetic stock at the beginning of the development period comprised 100,000 Awassi ewes, 2,000 Merino ewes and 200 Finn cross ewes. The minimum price for hired labour was set at $ 5,000 per person-year. In the 'standard' run, prices for tradable products per kg were set at $ 0.20 for wheat exported out of the region, $ 0.17 for imported concentrates, $ 0.60 for N and P fertilizer, $ 1.7 for mutton on the export market and $ 2.2 for mutton on the local market.

7.4.2. *The goals*

The potentials of the interactive multiple-goal linear programming technique are best exploited, when the number of goals specified in the model is initially set at a high number. That provides optimum flexibility and allows one to keep the technically feasible development pathways as open as possible (van Eijk *et al.*, 1986). In the example presented here, the following goals have been defined:
- Development aid, *i.e.* money for imported capital goods that is made available to the region at no cost and does not have to be repaid in the course of development. When this goal is set to zero, the possibilities for the region to develop on its own can be investigated. On the other hand, allowing infusion of limited amounts of development aid could indicate the scope for stimulating effects of such investments.
- Use of imported concentrates. Minimization of this goal could serve the national government, especially if the costs of imported concentrates include

182

a substantial proportion of foreign exchange. However, that may increase the need for import of fertilizers for production of feed locally, because nutrient availability is a major constraint for production (Chapter 3).
- Consumption by the local population, defined as income before taxes less investment. Maximizing consumption over time is a goal of the local settlers. Consumption may also be a goal of the national government in as far as it serves as a tax base.
- Employment. Maximum employment opportunities are a common development goal. In this respect a distinction is made between the local settlers for whom no alternative employment opportunities are assumed, and migrant, hired labour that has to be paid. Restricting the number of settlers and allowing unrestricted hiring and firing of migrant labour would be in the interest of the local population, but increasing the population without creating unemployment may be the goal of the national government or a settlement agency.
- Traditional system conservation. This goal of maintaining traditional production techniques and animal breeds in the region may be motivated by arguments relating to maintenance of environmental diversity and preserving animal and plant genetic stock. For the purpose of illustration, such extensive production techniques can be based on the local Awassi sheep breed, grazing yearlong without fencing, and with roughage as the main supplementary feed.

7.4.3. *The policy views*

To illustrate the capabilities of the multiple-goal linear programming technique for the purpose of agricultural development planning, three policy alternatives were formulated:
- The view of the present settlers, based on free enterprise. Their goals can be summarised as: restricting the number of settlers, maximize consumptive income for the present population, unlimited hiring and firing of migrant labour, no more than a minimum area with traditional production techniques, and no limitations on the use of imported concentrates.
- The view of a settlement agency, whose goals include: maximization of the number of settlers, with an income at least equal to that of the price of hired labour, low unemployment, no specific requirements with respect to the area under traditional production techniques and no limitations on the import of concentrate feed.
- The view of a conservationist group. Their goals include minimum use of imported concentrates, a large area under traditional production techniques, a reasonable number of settlers in the region, with an equitable income, and limited use of migrant labour.

Table 7–4.
Upper and lower limits over a fifteen-year period for three goal variables as determined in the second iteration cycle of MGLP.

Goal maximized	Values of goal variables		
	Cons. income 10^6 \$	Employment 10^2 pers.-year	Traditional systems 10^3 ha
Cons. income	197 a)	135	112
Employment	50 c)	192 a)	100 b)
Traditional systems	50 c)	58 b)	742 a)

a) upper limit; b) lower limit; c) lower bound.

7.4.4. *Results*

The extreme attainable values of the goal variables were determined one by one, while setting the requirements on the other goals either at their lowest physical limit or at a very modest level. This procedure provides for each of the goal variables the highest value that can possibly be attained and sets also the lowest value that need be accepted. After this first iteration cycle it was decided that the region had sufficient potential to develop without external financial resources, hence development aid was not further investigated, and set to zero. Additionally, consumptive income, necessary for the local population over the planning horizon, was estimated to be at least \$ 50 million, derived from the average number of settlers over the period and the price of hired labour. In the subsequent iteration cycle, three goals, reflecting the different policy views were further examined: consumptive income, employment and the area under traditional production techniques. The results of those optimization runs are given in Table 7–4. When consumptive income was maximized, a value of \$ 197 million could be achieved, which under the prevailing economic conditions thus represents the maximum. In the runs where the two other goals were maximized, consumptive income drops to \$ 50 million only, hence the lower bound set in the model. When employment is maximized, a total of 19,200 person-years can be achieved over the total planning horizon of 15 years, whereas employment only reaches 5,800 person-years in the situation where the area under traditional production techniques is maximized. When the area under traditional production techniques is maximized, a total of 742,000 ha-years can be exploited in that way over the planning horizon of 15 years, out of a physical maximum of 750,000 ha-years. The requirement of a minimum consumptive income of 50 million dollars prevents complete utilization of the land under traditional production techniques. In the run where employment is maximized, only 100,000 ha-years under traditional production techniques are chosen, which defines the lower value that needs to be accepted.

The results in Table 7–4 define the solution space for the region in the absence of outside financial aid. For the further analysis it was assumed that the

Table 7–5.
Upper and lower limits over a fifteen-year-period for two goal variables as determined in the third iteration cycle of MGLP, serving the policy of the conservationist group.

Goal maximized	Values of goal variables		
	Cons. income 10^6 \$	Employment 10^2 pers.-year	Traditional systems 10^3 ha
Cons. income	144 a)	113 b)	600 c)
Employment	90 b)	131 a)	600 c)

a) upper limit; b) lower limit; c) lower bound.

conservationist interest group considered both the consumptive income of \$ 50 million and the employment of 5,800 person-years too low, but was prepared to accept a decrease in the area under traditional production techniques to 80 percent of the potential. That requirement was, therefore, introduced in the model as the lower bound and consumptive income and employment were optimized again. The results of this iteration round (Table 7–5) show that under these conditions consumptive income ranges between \$ 90 and 144 million, and employment between 11,300 and 13,100 person-years. The conservationist group, reflecting on these results decided that employment should at least be 12,500 person-years. Setting that value as the lower bound in the model, and optimizing again for consumptive income resulted in a value of \$ 135 million. Hence, the conservationist's 'world' provides the region with a consumptive income of \$ 135 million and a total employment of 12,500 person-years, which implies an annual income before taxes of about double that of hired labour. These results could of course be challenged by one of the other interest groups, who could argue that such a high proportion of traditional production techniques in the land use of the region involves too high a cost in terms of consumptive income, or employment. A further analysis can then be carried out, in which the consequences of these views can be analysed quantitatively, so that gain in consumptive income and/or employment can be expressed in terms of the sacrifices required in reduced area under traditional production techniques.

The consequences of various policy options for selection of production techniques and the output of the region are now elaborated. The policy options examined are basically identical to those defined before, but some additional bounds on the goal variables were defined: for the settlement agency and the settlers a lower bound of 75,000 ha under traditional production techniques was introduced, for the conservationist interest group import of concentrate feed was not allowed, and for the settlement agency hire of migrant labour was restricted to 200 person-years per year.

The results are presented in Table 7–6, showing that total employment varies between 10,600 and 15,000 person-years over the 15-year period. The settler's policy favours restriction of the total number of settlers, and as a consequence,

Table 7-6.
Summary of results of the MGLP model for three policy views, under the limiting goal variable values imposed.

Interest group	Conservationist interest group	Settlement agency	Settlers	
Employment	13,000	15,000	10,000	p.yr
Settlers	11,740	14,800	6,000	p.yr
Linear growth rate settlers	43	74	0	p.yr.yr^{-1}
Hired labour	1,260	200	4,600	p.yr
Consumptive income	121	196	169	10^6 $
Consumptive income per settler	11	14	28	10^3 $.yr^{-1}
Average results for years 7-12 (expressed on a per year basis)				
Mutton sale local market	5.5	9.7	9.1	10^6 kg
Mutton sale export	0	1.5	0.7	10^6 kg
Wheat sale export	0	11.3	15.2	10^6 kg
Imported concentrates	0	33.6	30.1	10^6 kg
Intermediate use wheat	13.4	7.7	0.2	10^6 kg
Roughage (straw) production	8.9	11.2	8.0	10^6 kg
N fertilizer use	1.6	0.65	0.6	10^6 kg
P fertilizer use	0.24	0.87	0.6	10^6 kg
Area wheat/fallow	0	0	0	ha
Area wheat/wheat	8,400	3,700	100	ha
Area wheat/legume	851	15,100	16,900	ha
Total area wheat	9,251	18,800	17,000	ha
Area extensive production techniques	60	12	8.3	10^3 ha
Total concentrate use (imported, wheat grain, legume straw)	104	359	327	kg ewe^{-1}
Investments	1.9	2.1	4.3	10^6 $

it also requires the most hired labour. Total consumptive income is intermediate in this policy view, but as the number of settlers is low, the *per capita* income per settler is by far the highest in this scenario.

The scenario set by the settlement agency leads to the highest total employment, and, not surprisingly, to the highest growth rate of settlers in the region. Total consumptive income is also highest here, but because of the high rate of increase of settlers in the region, *per capita* income is only half of that achieved under the scenario favoured by the local population. However, before taxes it is still three times higher than the cost of hired labour.

Implementation of the development plan of the conservationist group leads to an intermediate level of employment, with a lower increase in population growth in the region. Both, total consumptive income and income *per capita* are

Table 7–7.
Sheep husbandry systems selected by the model expressed in ewe equivalents under the three scenario's, averages of year 7–12 (see also Tables 5–1 and 5–2).

Code	Environmentalist group	Settlement agency	Settlers
Group V, rangeland, unfenced, roughage			
impr-yearl-norm,G=0.7	4,600	0	0
impr-yearl-norm,G=0.9	22,700	13,000	8,200
GMM -defer-earl,G=1.6	0	17,300	320
total group V	27,300	30,300	8,520
Group VI, rangeland, fenced, concentrates			
impr-yearl-norm,G=0.9	0	700	4,100
GMM -defer-norm,G=1.4	0	0	5,700
GMM -defer-earl,G=1.6	0	0	13,200
total group VI	0	700	23,000
Group I, cultivable land, unfenced, roughage			
impr-yearl-norm,G=0.9	84,800	0	0
impr-green-norm,G=1.0	10,000	0	0
GMM -green-norm,G=1.7	15,100	0	0
GMM -green-norm,G=2.0	8	0	0
total group I	109,908	0	0
Group II, cultivable land, unfenced, concentrates			
impr-yearl-norm,G=0.9	0	16,400	3,300
impr-green-norm,G=1.0	0	3,200	0
GMM -defer-earl,G=1.6	0	2,400	0
GMM -green-earl,G=1.9	0	21,200	0
total group II	0	43,200	3,300
Group IV, cultivable land, fenced, concentrates			
impr -yearl-norm,G=0.9	0	2,000	19,700
impr -green-norm,G=1.0	0	7,400	0
GMM -defer-norm,G=1.4	0	0	11,300
GMM -defer-earl,G=1.6	0	3,100	3,900
GMM -green-norm,G=1.7	0	0	400
GMM -green-earl,G=1.9	0	0	21,700
FinnX-green-earl,G=2.4	0	0	3,300
total group IV	0	12,500	63,900
Group VII, feedlot, roughage			
GMM -norm,G=1.7	83	0	0
FinnX-norm,G=2.0	2	0	0
total group VII	85	0	0
Group VIII, feedlot, concentrates			
GMM -earl,G=1.9	0	27,950	10,400
FinnX-earl,G=2.4	0	27,150	23,800
total group VIII		55,100	34,200
Total number of sheep of various breeds			
Awassi	0	0	0
Improved Awassi	122,100	42,700	35,300
German Mutton Merino	15,191	71,950	66,520
Finn Cross	2	27,150	27,100
total number of sheep	137,293	141,800	128,920

the lowest under this scenario, but the latter is still twice as high as that assumed for hired labour.

Technology selection has been analysed for the period from 7–12 years after the start of the development period. This period has been chosen to avoid distortions due to initialization and 'termination' artefacts, as explained in the preceding section.

Annual mutton production in the region is highest in the scenario advocated by the settlement agency, *i.e.* 11.2 million kg. The mutton produced for export is valued at a lower price than that for the local market. In the settlement agency's scenario only about 13 percent of the total production has to be exported. The lowest mutton production results when the policy scenario of the conservationist group is implemented, as can be expected under the condition of import restrictions on concentrate feed. That constraint is, however, partially alleviated by increased use of locally produced wheat grain as a substitute, a practice that is less favoured in the other two scenario's, as the price of imported concentrate is lower than that of exported grain. The consequence of the increased grain production for animal feed is, that in the conservationist's scenario the use of nitrogen fertilizer is relatively high. However, this is considered acceptable because the efficiency of nitrogen in semi-arid regions is high (Chapter 3). Since there are hardly any feedlots in this scenario (Table 7–8), there is no farmyard manure available to reduce the use of industrial fertilizer. More farmyard manure is produced in the other scenarios, but proper storage and use of farmyard manure for wheat fertilization is too expensive. It is treated here as waste, *i.e.* as long as government regulations allow it.

With regard to arable farming, continuous wheat cultivation is favoured in the scenario of the conservationist, while the wheat/legume rotation is heavily favoured in the two other scenario's. The wheat/fallow rotation is never selected. It indeed is a relatively uncommon practice in the region, but it is sometimes practiced for phytosanitary reasons. The wheat/legume rotation is favoured in the scenario's of the settlers and the settlement agency, as it reduces the dependence on expensive imported concentrate, and hence increases consumptive income. That is an explicit goal of the settlers, and an implicit goal of the settlement agency, as a minimum livelihood for increased numbers of settlers is aimed at.

Definition of system parameters:

Sheep breed: impr = Awassi under improved management;
    GMM = German Mutton Merino; FinnX = Finncross.
Grazing system: yearl = yearlong; green = green season only
    (includes early dry season when quality of pasture is still high);
    defer = yearlong except for deferment during transitional period after opening winter rains.
Weaning system: norm = normal weaning; earl = early weaning
G = net lambing rate: lambs weaned per ewe per year.

The lowest investments are necessary in the scenario for the conservationist group, mainly because the requirement for a large area under traditional production techniques reduces the need for expensive fencing. In addition, the total wheat area is smaller, so that investments for capital assets associated with wheat cultivation are not necessary. The highest investments are required in the scenario favoured by the settlers. These needs are associated with extensive use of labour-saving fences, building of physical structures and equipment on both rangeland and cultivable land.

The various sheep husbandry systems selected by the model in the three scenario's are summarized in Table 7-7, again as averages for the period 7–12 years. In the scenario favoured by the conservationist group, Awassi under improved management is by far the most selected sheep breed. All sheep husbandry techniques selected are on unfenced land, and feedlot operations are hardly selected. In the scenario's of the settlers and the settlement agency, sheep breed selection is very similar, with slightly over half the animals being Merinos, the remainder being equally split between Awassi under improved management and Finn-cross ewes.

The main difference between the development scenario of the settlers and that of the settlement agency is that in the former most of the sheep production techniques are practiced on fenced land, while in the latter production techniques on unfenced land prevail. Evidently, that is a consequence of the employment requirements imposed on the model, as herding on unfenced land requires substantially more labour. Also feedlot operations are selected to a larger extent by the settlement agency, as these are relatively labour-intensive too.

7.5. Discussion

The interactive multiple-goal linear programming technique can help planners who have to analyse agricultural development possibilities on a regional basis. It is a tool they can use to choose between different feasible development pathways under a wide range of technical and socio-economic conditions. The model can incorporate a large amount of general and local knowledge on actual and potential production techniques, regional physical and socio-economic resources and constraints and prices of inputs and outputs. This makes it possible to conduct an analysis of a dynamic planning process that explicitly takes into account a large number of technical possibilities and regional interests. The model has a strong technical base, in which socio-economic constraints, like ownership of the means of production, distribution of income, and uncertain economic behaviour are not considered. This approach avoids the very complex problems associated with these aspects of development and allows for full, untrammelled analysis of a wide range of technically feasible development pathways including innovative, unexpected ones. As a consequence of this approach, the results may raise too optimistic expectations for development in a region. For instance, a desirable development may require common use of scarce resources, which may only be possible through policy measures aiming at promoting or even forcing cooperation, and legislation may be required to regulate such cooperation. Here the political dimension comes into play and the analysis can help to define the benefits and costs that could motivate desirable change. The socio-economic dimension is then also clearly presented with its 'margins for policy' (de Wit *et al.*, 1988). It is, therefore, not only a convenience, but important in principle, to separate as strictly as possible the technical coefficients of the present and possible alternative agricultural production techniques on the one hand and the socio-economic environment in which they are to be implemented on the other. The consequences of a proposed development pathway can be evaluated in terms of economic viability and social acceptability. If some of the socio-economic constraints then appear insurmountable over the planning horizon, it may be necessary to change some of the goal restrictions or to introduce new constraints and repeat the planning exercise until a workable plan is attained.

By quantifying the implications of implementing a certain viewpoint, such a model can also serve as a basis for negotiations between various interest groups in a region. That may lead to a compromise and so provide a broader base for the proposed regional development plan. In this way the results of the model can be used to improve communication between planners and policy makers and between policy makers and various interest groups in a region. This can help to smooth the way to a more balanced development in which the interests of all the parties that are involved are taken into account.

The validity of the model results depends on many factors. These include the accuracy of the technical coefficients in the input-output model, the proper definition and quantification of the goal variables and the degree to which the technical and socio-economic possibilities can be treated separately. Technical coefficients for present production techniques often can be determined with sufficient accuracy on the basis of the general body of knowledge in a region. However, there is a danger that the technical coefficients of alternative production techniques are estimated too optimistically by their advocates and critical evaluation is, therefore, necessary. Equally important are the goal variables, which are often difficult to identify, difficult to translate into terms relevant for the model and difficult to quantify.

The separation of technical and socio-economic aspects, which is an essential feature of the approach, makes it possible to distinguish between the technical possibilities in a region and the elusive behavioural factors that often strongly influence actual development policies. As a result, the analysis can help to define political issues more meaningfully, and so '... grant unto Caesar what is Caesar's', while pointing to feasible pathways for development. In this way, the discussions about development options can be conducted in more tangible terms, and so make communication between different stakeholders easier and more to the point.

The results of such an analysis are only an intermediate step in formulating actual development policies. The model results indicate the requirements for promotion of desired development options, but they do not answer the question as to what policy measures are necessary to actually bring them about. A post-model analysis is necessary to translate the requirements for external inputs, investments, education, research, etc. into practical actions. In this analysis the social acceptability of the proposed developments should be given special, careful consideration.

7.6. Conclusions

The interactive multiple-goal linear programming technique for analysis of rural development possibilities can perform a number of important functions:
- Identification of consistent, technically feasible development pathways for a region, that promote a combination of goal values considered most acceptable to the stakeholders in the sector.
- Evaluation of the cost of achieving full realization of one goal in terms of the sacrifices that are imposed by regional constraints on other goals.
- Translation of the selected combination of goal values into a set of production techniques that are required for their achievement.

These can be expressed as requirements for land reclamation, imports of means of production, export of products, credit facilities, education, etc. This analysis can provide a choice of technically feasible options that can serve as a well-constructed objective basis for further socio-economic and policy analysis.

Appendix

Model activities and constraints by categories

Activities
I.   Production
     A. Animal production systems (108 technologies)   head
     B. Crops
         Grain – grain
         Grain – fallow
         Grain – legume

II. Capital formation and/or use
     C. Buildings and equipment $
         Animal buildings and machinery on range land
         Fences on range land
         Animal buildings and machinery on cultivable land
         Fences on cultivable land
         Artificial lamb rearing units
         Equipment for crop cultivation
     D. Retention of hoggets for reproduction   head
         Awassi

Improved Awassi
German Meat Merino
Finn Cross

III.	Intermediate products	
	E. Animals	head
	Awassi lambs	
	Improved Awassi lambs	
	German Meat Merino lambs	
	Finn Cross lambs	
	F. Crop products	kg
	Grain for obligatory concentrate feed	
	Grain or legume pasture for replaceable concentrate feed	
	Straw	
IV.	Hired labour	
	G. Hired labour	person-year
V.	Trade across the regional border	
	H. Purchases	
	Obligatory concentrate feed	kg
	Replaceable concentrate feed	kg
	Nitrogen fertilizer	kg
	Phosphorus fertilizer	kg
	Other purchased inputs	$
	I. Sales	
	Mutton for the domestic market	
	Mutton for the export market	
	Grain	
VI.	Cash flows	$
	J. Investment	
	K. Aid	
	L. Borrowing	
	M. Consumption	

Constraints

1.	Natural resources	
	Range area	ha
	Cultivable area	ha
	Water for livestock	m³
2.	Capital stock	$
	Animal buildings and machinery on range land	
	Fences on range land	
	Animal buildings and machinery on cultivable land	
	Fences on cultivable land	
	Artificial lamb rearing units	
	Equipment for crop cultivation	
3.	Labour force	person-year
4.	Animal feed	FU
	Obligatory concentrate feed	

	Replaceable concentrate feed	
	Roughage	
5.	Fertilizer	kg
	Nitrogen fertilizer	
	Phosphorus fertilizer	
6.	Other inputs	$
7.	Animals-ewes	head
	Awassi	
	German Meat Merino	
	Finn Cross	
8.	Animals-hoggets	head
	Awassi	
	Improved Awassi	
	German Meat Merino	
	Finn Cross	
9.	Mutton	kg
	Lamb and culls for export	
	Lamb and culls for the domestic market	
10.	Main crop products	kg
	Grain	
11.	Secondary crop products	ha
	Straw	
12.	Investment capital	$
13.	Revenue	$

8. Summary

C.T. DE WIT

8.1. Introduction

Is water the dominant constraint to primary production in semi-arid regions? That question sparked the study described in this book. At an early stage it became clear that nutrient deficiency was at least as important, and even more so in average to good rainfall years. Consequently, the study was expanded to include the effects of improved nutrient availability on pasture growth and crop yields in the semi-arid environment of the northern Negev. This led to the development and analysis of crop and sheep husbandry systems that were designed to utilize the arable land and improved pastures more efficiently. The choice of agro-pastoral systems whereby primary and secondary production can be integrated in a given region was the next subject to be broached and led to the problem of considering simultaneously relevant agro-ecological as well as socio-economic constraints that apply to regional agricultural development.

As conditions change and especially when active development is pursued, far reaching and rapid changes in husbandry systems become the order of the day as old-established traditional systems and even newer systems become obsolete. Innovative adaptations have to be identified, usually under conditions of increasing uncertainty with regard to the future. To analyse the possibilities for regional development, a three-step approach was elaborated. First, the feasibility and robustness of selected innovative techniques at the farm level were investigated in relation to variabilities in weather and prices, then a matrix of production techniques for a region was formulated in terms of their physical inputs and outputs, and finally this matrix was embedded into a dynamic multiple-goal linear programming model that facilitated the exploration of different development pathways in dependence of socio-economic aspirations and constraints.

8.2. Primary production

Vegetation that grows in the rainy winter of a semi-arid mediterranean climate must be able to survive the long, hot and dry summer. Apart from deep-rooting dwarf shrubs, annuals that enter the dry period in the form of seeds are well

193

Th. Alberda et al. (eds.), Food from Dry Lands, 193–200.
© 1992 *Kluwer Academic Publishers.*

adapted to this situation. Most of these have an extended and flexible period of seed formation, so that even in years when the first rains are late or when late rains fail, enough seeds are usually formed for the species to survive. These species, and in particular those that do not have characteristics that are objectionable to foraging ruminants (like thistles, sharp awns or noxious constituents), provide a good basis for animal production.

The quality of these annuals as animal feed is characterized mainly by their protein content and digestibility. These depend to a large extent on the growing conditions. While phosphorus may be deficient in some soils in the semi-arid mediterranean regions, nitrogen is always in short supply. This becomes obvious in wet years when nitrogen fertilizer must be supplied in considerable amounts to achieve a situation in which water is the constraining factor. If then no fertilizer is added, the limited amount of nitrogen available is diluted to such an extent that the protein content in the material approaches its minimum value. In dry years, when water is the limiting factor and biomass production is low, the vegetation has a high protein content even at maturity, because the amount of nitrogen available from natural sources is diluted much less. These observations were confirmed by experimental results that also served to verify dynamic simulation models of annual pasture growth under conditions of limited water availability. These analyses also showed that the potential production rate of native pastures, particularly during the vegetative period, was identical to that of pastures of cultivated species or small grain crops. The differences in final yield were the result of different phenological patterns and growing periods as well as differences in allocation of assimilates between the various plant organs.

Comparison of the calculated and actual production levels over a thirteen-year period showed that in most of the years the available amount of water could have indeed supported a considerably higher production. Hence, the water resource, although scarce, is very much under-utilized in semi-arid regions. Fertilizer application can thus increase the average yield of native pastures considerably but, as yields in drought years are not affected, it also increases the year-to-year variability.

Since seasonal rainfall cannot be predicted, expensive fertilizer applied in years that turn out to be dry can be a financial burden. Fortunately, under semi-arid conditions, and particularly in a dry year, losses of nitrogen by leaching or denitrification are small. Consequently, most of the applied nitrogen that is not recovered from the soil by the vegetation in the year of application because of lack of water, is available for vegetation growth in the following year. With an appropriate fertilizer application strategy, fertilizer use efficiency can be maintained at a high level so that, eventually, most of the fertilizer nitrogen applied to pasture ends up as crude protein that can be utilized by grazing livestock. Phosphorus is in general not lacking in mediterranean pastures, and where it is, its application poses no specific problems beyond those of an economic nature.

In small grains, nitrogen deficiency limits growth and dry matter production in the same way but the effect on grain yield involves a major complication: a liberal supply of nitrogen may stimulate early growth to such an extent, that the water in the soil is exhausted before seed filling starts. When this happens, grain yield is lower as a consequence of nitrogen application.

Both natural and fertilized native pastures produce far more seed than is required for regeneration of the pasture in the following year. Most of these seeds are 'harvested' by granivores, mainly ants but also rodents and birds. As a rule, the remaining seed stock, usually protected in refuges provided by the soil and residual vegetation, is sufficient to ensure establishment of the vegetation in the subsequent year. The combination of fertilizer application and green season grazing radically changed the species composition of the pasture. While this change did not endanger continuity of exploitation, it did reduce species diversity and the species that came to dominate the pasture were not necessarily the most valuable. Nevertheless, both primary and secondary production were dramatically increased and viable measures are available to improve the vegetation of pastures that became dominated by less desirable pasture species.

Utilization of the vigorous pasture growth achieved by fertilizer application requires high stocking densities. The average growth rate of fertilized pasture during its grand period of growth is about 100 kg dry matter per ha per day, except in drought years. Since the maximum intake per ewe is normally less than 3 kg per day, very high stocking densities would be needed for defoliation to affect production seriously. However, in the early growing season during the first two months after germination, growth rates are much lower because of incomplete soil cover and low temperatures. At that stage, heavy grazing of the highly nutritious young vegetation can significantly reduce seasonal production and the tradeoff between immediate benefit and total pasture production must be considered, particularly when other feed sources may be scarce. Consequently, the subject of grazing deferment at the beginning of the growing season was analysed in detail.

The same considerations apply to very early grazing of wheat. Here not only total dry matter production but also grain yield is important. It could be shown that this was hardly affected, even by heavy grazing, during a period of up to three weeks between canopy closure and the start of stem elongation. When pasture is scarce, particularly during the early part of the season, this extra bite can be important and is another good reason for integrating arable and pastoral land use.

8.3. Secondary production

Sheep are better adapted to a semi-arid annual vegetation type than cattle. Their shorter reproductive cycle facilitates adaptation to the strong seasonality of primary production and their early puberty makes it easier to adjust flock size to annual climatic or economic fluctuations. If pasture production is low, feed requirements for lactation can be reduced by early weaning of lambs and if pasture production is high, out of season lambing is possible. Adjustments to different production levels can also be realized by appropriate choice of sheep breeds and management systems. In areas where productivity of natural pasture is low, hardy, well adapted native breeds with moderate prolificacy, like the Awassi, predominate. Traditional systems, based on this breed, are characterized by single-lamb litters and by a lambing frequency of once per year in good years and lower frequencies during periods of extended nutritional stress. Intensification of production beyond a modest improvement, either by the use of supplementary feed or by pasture improvement, soon requires the use of breeds with higher prolificacy. In Israel, at first the emphasis was on improved management of the Awassi breed, but at a later stage German Mutton Merino sheep were imported. The introduction of rams of the prolific Finnish land race and Romanov breeds resulted in crosses that produced twins and triplets, so pushing the average prolificacy up to 1.65 lambs per ewe per lambing. With multiple breeding of up to 3 lambings every two years, the lamb crop can be increased to more than 2 lambs per ewe. However, because of the greater dependence on supplementary feeds and susceptibility of the crosses to diseases, more care is required and as a result, their full production potential can be attained only in intensively managed feedlot systems.

In improved grazing environments the emphasis shifts from per animal production to per unit area production and this is closely related to stocking rate. The experimental work centered, therefore, around the interactions between stocking rate and herbage production, the fraction consumed in the growing season and the daily herbage intake of individual sheep. In general, there is a diminishing return in the production per sheep as stocking rate increases. However, sheep production per unit area continues to increase with increasing stocking rates well beyond the point where per sheep production begins to decrease. In intensive agro-pastoral systems sheep are not only supplemented because of the low quality of the dry pasture during part of the year, but also when sheep performance declines due to overstocking. However, the added cost may significantly reduce the economic efficiency of the system. In addition, annual variation in pasture production, particularly when fertilizer is added to the pasture, is a complicating factor. If under these conditions flock size is fixed at a level that prevents overexploitation in dry years, large amounts of good quality grass are wasted in favourable years. Fertilization is then hardly attractive. On the other hand, if the grass is fertilized and flock size is adjusted to a level that leads to utilization of most of the grass in more favourable years, feed shortages occur in drier years. These can be overcome only to a limited degree by increased culling and sale of unproductive animals and so it may be necessary to resort to more supplementary feeding. Fertilization of natural pastures in semi-arid regions is, therefore, invariably accompanied by stocking rates adapted to the more favourable years, and supplementary feeding, even during the growing season, in the drier years. The level of intensification justified by the prevailing bio-physical, climatic and socio-economic conditions must be carefully considered, and the analysis can involve a complex array of factors. Some of the models developed as part of this study were designed to facilitate such an analysis.

8.4. The matrix of production techniques

A matrix of production techniques was used to construct a multiple-goal programming model that contains the technical coefficients that specify the inputs and outputs of all the appropriate agro-pastoral systems that are known to be relevant to semi-arid regions with winter rains. The production techniques are essentially the same in different regions, but quantitative differences exist, due to differences in soil and climate. The technical coefficients of the more extensive pastoral systems and of feedlot operations can be derived from a common base of knowledge and experience. However, systems characterized by a more intensive land use, are less well understood and their technical coefficients can only be derived from a sometimes speculative combination of the existing base of knowledge with new information on primary and secondary production, as presented in this book.

To cover the whole spectrum of options, a systematic approach has been developed, in which the systems are distinguished on the basis of breed, land unit, fencing, fertilizer regime, grazing regime and the use of arable crops and legumes. The systems also differ in their use of labour, transport, buildings, watering points, veterinary services and the level of technological know-how that is required. Apart from some mutton and wool, the main output of each technology is lamb meat. All inputs and outputs are long-term averages defined in physical terms with the ewe as a common denominator.

The procedure for deriving the technical coefficients is target oriented in the sense that the production level of each system is pre-determined and the inputs necessary to achieve that target are subsequently derived. In this way, more than 100 production techniques were generated. This number may seem unrealistically high, but it should be emphasised that when the study is at a stage where regional constraints, aspirations and prices have not yet been defined, only those systems that use more of each input per unit output than any other system can be safely discarded as inefficient.

In general terms the matrix of techniques divides into three levels of intensity, defined by breed, nutrition and prolificacy:
- extensive systems, in which the local fat-tail Awassi breed with net lambing rates from 0.5 to 1.0 lambs per ewe per year is fed on pastures with no or limited use of fertilizer;
- intensive systems, in which Merino crosses with net lambing rates from 1.0 to 1.5 lambs per ewe per year are fed on fertilized pastures and concentrate supplements;
- highly intensive systems, in which Finn crosses with net lambing rates from 1.6 to 2.4 lambs per ewe per year are mainly fed on concentrate feed.

8.5. Management under uncertainty

The matrix of production techniques is the basis for determining long-term strategic decisions in a development context. However, many of the systems defined in the matrix are innovative and there is, therefore, no practical experience on which to base an estimate of their viability and robustness under the regional conditions. Consequently, it was necessary to investigate some of these more carefully in the context of a management environment where short-term, tactical decisions have to be taken in response to the current state of the system and the expected short-term performance. These include decisions on grazing schedules, grazing deferment of paddocks, lamb rearing, supplementary feeding of lambs and cutting of wheat for hay.

For this purpose, criteria for short-term decisions were formulated into appropriate optimization algorithms which were then used to analyse the response to management of a family of sheep and agro-pastoral systems that represented an important group of innovative systems. In this optimization model, a daily time interval for the biological routines and a five-day time interval for the management routines are used. Each run of the model refers to a 21-year simulation, using weather data for the northern Negev from 1962–1983.

The results provide a framework for analysing the relative importance of various management options, the sensitivity of system output to management and the range of economic and biological efficiencies under weather uncertainty and, hence, the feasibility and robustness of the system under consideration.

The results indicate that the pastoral systems under consideration 'look after themselves', in the sense that major differences in economic scenarios only have a small effect on most of the important technical decisions. This is characteristic for complex systems with strong negative feedbacks between components and adds confidence to the applicability of the innovative production techniques that were selected by the interactive multiple-goal planning process.

8.6. Regional development

Development involves the introduction of innovative technologies that use the local and available external resources more efficiently to serve the goals and aspirations of those that have a stake in the sector. In the present context, the possible development pathways within the physical and socio-economic constraints of the region are identified and analysed as an interactive dynamic linear programming problem, based on the matrix of production techniques as discussed above. The production process converts inputs into usable outputs that can be used within the region or traded over the region border. Some inputs, like labour, are available within the region. Others, like concentrates, fertilizers, fencing material and means of transport, have to be bought from outside the region. The main outputs from the agro-pastoral sector are meat and grain. In its most simple form, the economic environment can then be defined by a set of relative prices for these tradable goods.

The constraints relevant to the agro-pastoral sector in a region include the area and quality of the available land, the size of the population that depends on it, the initial endowment of capital goods and the number and breeds of livestock. The rate of accumulation and obsolescence of capital goods as well as the change in flock size and breed composition are slow, and so must be placed within an appropriate planning horizon of at least 15 years.

To evaluate the performance of the model, a regional scenario was defined with boundary conditions similar to those of the northern Negev and optimizations were performed to examine in more detail the course of investments, the generation of revenue for saving and consumption and the dynamics of sheep breed composition in comparison to actual developments. Subsequently, the ability of the model to reflect differences in socio-economic conditions in a realistic way was investigated by analysing developments in three widely differing socio-economic environments that broadly represent the northern Negev in Israel, a region in the western Australian wheat belt and the coastal region of the western Egyptian desert.

The ultimate purpose of these analyses was not so much to find some optimal development path to satisfy a single goal, like the generation of revenue for consumption, but to explore to what extent available production techniques within the physical and socio-economic constraints of the region, can meet various and partly conflicting demands by, for instance, the government, development agencies, various interest groups and the local population. A settlement agency may want to increase the number of settlers, the local population may emphasize their own gainful employment and their own consumptive income, whereas conservationist groups may want to retain extensive pastoral systems and put limits on the use of concentrates and fertilizers. Also, the central government may be concerned about the contribution of the sector to the balance of payment and the World Bank may emphasize control over the inflow and outflow of capital.

An interactive multiple-goal planning procedure was used to analyse perspectives for development in a series of iterations. First limiting values are specified for each of the goals. These are the absolute minimum or maximum values for each goal that are acceptable to the policy maker. Then the maximum attainable value for each goal is calculated while all the other goals are held at their minimum values. In the following iterations, goal restrictions are tightened one by one as the other minimum goal attainment values are improved. The aspirations and the interests of the interactively operating policy maker determine which goals will be tightened and to what degree. In this way, the costs of realizing one goal are expressed in terms of what must be sacrificed on the others. Persons with different aspirations and interests are bound to end up in different corners of the original solution space. Hence, the analyses help to explore perspectives for development and to evaluate alternative options that will affect the resources and technical structure of the sector. The results indicate which technical developments are needed to satisfy the goals and aspirations of the policy makers, but they do not determine what policy measures would be needed to achieve them.

In this way, the decision maker can compare different scenarios for a given region and study their interrelations and consequences. While the selected development pathway may not in fact be implemented in the end, at least the options will have been very thoroughly analysed and some illusions and pitfalls will have been foreseen and, hopefully, avoided. This approach can thus provide a formal bridge between the biophysical and the socio-economic elements of agro-pastoral development and in that way allow for more effective interdisciplinary application of scientific knowledge to arid zone development.

References

Aase, J.K., 1976. Relationship between leaf area and dry matter in winter wheat. Agron. J. 70, 563–565.

Amir, D. and R. Volcani, 1965. The sexual relation of the Awassi fat-tailed ewe. J. agric. Sci. (Camb.) 64, 83–85.

Amir, J., A. Vananu, H. Krikun, D. Orion, Y. Penuel, Y. Satki and A. Lerner, 1982. Long-term experiments on dry farming in the Negev desert, 1974–1980. C. Cereal nitrogen economy in a semi-arid region. Hassadeh 62, 570–592. (Hebrew, with English summary.)

Angus, J.F. and M.W. Moncur, 1977. Water stress and phenology in wheat. Aust. J. agric. Res. 28, 177–181.

ARC, 1965. The nutrient requirements of farm livestock. 2. Ruminants. HMSO London. 37 pp.

ARC, 1980. The nutrient requirements of ruminant livestock. Commonwealth Agric. Bur., Farnham Royal, UK. 351 pp.

Arkley, R.J., 1963. Relationships between plant growth and transpiration. Hilgardia 34, 559–584.

Arnold, G.W., 1981. Grazing behaviour. In: F.W.H. Morley, ed. Grazing animals , Elsevier, Amsterdam. pp. 29–104.

Aspinall, D., 1961. The control of tillering in the barley plant. I. The pattern of tillering and its relation to nutrient supply. Aust. J. biol. Sci. 14, 493–503.

Baier, W. and G.W. Robertson, 1967. Estimating yield components of wheat from calculated soil moisture. Can. J. Plant Sci. 47, 617–630.

Balch, C.C. and R.C. Camping, 1962. Regulation of voluntary food intake in ruminants. Nutr. Abstr. Rev. 32, 669–686.

Barkai, D., 1979. The effect of limiting milk and early weaning on the feed intake and energy balance of lambs from birth to slaughter. M. Sc. Thesis, Fac. Agric., Hebrew University, Jerusalem. 32 pp.

Barkai, D., E. Eyal and R.W. Benjamin, 1981. The performance of weaned Merino-Finn lambs when fed concentrates *ad libitum* a sole diet or when fed as a supplement when grazing dry pastures. Internal Report, Dep. Range Sci., Agric. Res. Organ. (Bet Dagan). 12 pp.

Barnicoat, C.R., A.G. Logan and A.I. Grant, 1949a. Milk secretion studies with New Zealand Romney ewes. Parts I and II. J. agric. Sci. (Camb.) 39, 44–55.

Barnicoat, C.R., A.G. Logan and A.I. Grant, 1949b. Milk secretion studies with New Zealand Romney ewes. Parts III and IV. J. agric. Sci. (Camb.) 39, 237–248.

Bartholomew, W.V. and F.E. Clark, 1965. Soil nitrogen. Agronomy 10, Am. Soc. Agron., Madison, Wi. 615 pp.

Bauder, J.W. and A. Bauer, 1978. Irrigated alfalfa forage production on a well- drained sandy loam at Oakes, North Dakota 1972–1977. North Dakota Farm Res. 36, 19–24.

Benjamin, Y., 1983. A management model of a grassland management system under Israeli conditions. M. Phil. Thesis, Univ. Reading, Reading, UK. 162 pp.

Benjamin, Y. and D. Harel, 1983. Development of prices for meat from lambs and ewes. Hassadeh 63, 1141–1144. (Hebrew.)

Benjamin, R.W., E. Eyal, I. Noy-Meir and N.G. Seligman, 1976. The effect of sheep grazing on

Th. Alberda et al. (eds.), Food from Dry Lands, 201–211.

the grain yield and total dry matter production of wheat in an arid region. Hassadeh 57, 754–759. (Hebrew with English summary.)

Benjamin, R.W., M. Chen, A.A. Degen, N. Abdul Aziz and M.J. al Hadad, 1977. Estimation of the dry- and organic matter intake of young sheep grazing a dry mediterranean pasture, and their maintenance requirements. J. agric. Sci. (Camb.) 88, 513–520.

Benjamin, R.W., M. Chen, N.G. Seligman, D. Wallach and M.J. Al Hadad, 1978. Primary production of grazed annual pasture and of grazed wheat in a semi-arid region of Israel. Agric. Syst. 3, 205–220.

Benjamin, Y., J. Kali, E. Eyal, Y. Folman and D. Barkai, 1979. Utilization of maximum recycled materials – wheat straw and poultry litter – and minimal grain in feeding Awassi sheep for milk production. Hassadeh 59, 1237–1243 (Hebrew with English summary.)

Benjamin, R.W., A.A. Degen, A.R. abu Arafe and I. Noy-Meir, 1980. Lamb and ewe intake in relation to biomass availability. Internal Report, Dep. Range Sci., Agric. Res. Organ. (Bet Dagan). 23 pp.

Benjamin, R.W., Y. Benjamin and E. Eyal, 1981. The growth of lambs on natural pasture and sown legume pasture in the Negev in two drought years (1977/78 and 1978/79). Internal Report, Dep. Range Sci., Agric. Res. Organ. (Bet Dagan). 13 pp.

Benjamin, R.W., E. Eyal, I. Noy-Meir and N.G. Seligman, 1982. Intensive agropastoral systems at the Migda Experimental Farm in the northern Negev. Hassadeh 62, 2022–2026 (Hebrew with English summary.)

Biggar, J.W., 1984. Spatial variability of nitrogen in soils. In: D.R. Nielsen and J.G. MacDonald, Eds. Nitrogen in the environment. Vol. 1, 201–222.

Bircham and Hodgson, 1983. (No further details available.)

Black, C.A., 1966. Crop yields in relation to water supply and soil fertility. In: W.H. Pierre, D. Kirkham, J. Pesek and R. Shaw, Eds. Plant environment and efficient water use. ASA, SSSA, Madison, Wisc. pp. 177–206.

Black, J.L. and P.A. Kenney, 1984. Factors affecting diet selection by sheep. II. Height and density of pastures. Aust. J. agric. Res. 35, 565–578.

Blaxter, K.L. and A.W. Boyne, 1978. The estimation of the nutritive value of feeds as energy sources for ruminants and the derivation of feeding systems. J. agric. Sci. (Camb.) 90, 47–68.

Bolton, J.K. and R.H. Brown, 1980. Photosynthesis of grass species differing in carbon dioxide fixation pathways. V. Response of *Panicum maximum*, *Panicum milioides* and tall fescue (*Festuca arundinacea*) to nitrogen nutrition. Plant Physiol. 66, 97–100.

Bowden, L., 1979. Development of present dryland farming systems. In: A.E. Hall, G.H. Cannell and H.W. Lawton, Eds. Agriculture in semi-arid environments, Springer Verlag, Berlin. pp 45–72.

Breman, H., 1975. Maximum carrying capacity of Malian grasslands. In: ILCA. Evaluation and mapping of African rangelands. pp. 249–254.

Breman, H., 1982. La productivité des herbes pérennes et des arbres. In: F.W.T. Penning de Vries and A.M. Djitèye, Eds. La productivité des pâturages sahéliens. Une étude des sols, des végétations et de l'exploitation de cette ressource naturelle. Agric. Res. Rep. (Versl. landbouwkd. Onderz.) 918, Pudoc, Wageningen. pp. 284–296.

Briggs, L.J. and H.L. Shantz, 1913. The water requirements of plants. I. Investigations in the Great Plains in 1910 and 1911. U.S. Dept. Agr., Bur. of Plant Ind., Bull. 284. 49 pp.

Brouwer, R., 1962. Nutritive influences on the distribution of dry matter in the plant. Neth. J. agric. Sci. 10, 399-408.

Brouwer, R., 1963. Some aspects of the equilibrium between overground and underground plant parts. Jaarb. Inst. Biol. Scheik. Onderz. Landbouwgewassen 1962, 31–39.

Burg, P.F.J. van, 1962. Internal nitrogen balance, production of dry matter and ageing of herbage and grass. Agric. Res. Rep. (Versl. landbouwkd. Onderz.) 68.12, Pudoc, Wageningen, The Netherlands. 131 pp.

Carter, E.D., 1981. A review of the existing and potential role of legumes in farming systems of the Near East and North African region. Report to ICARDA, Waite Agric. Res. Inst., Glen

Osmond, South Australia. 120 pp.

Cissé, A.M. and H. Breman, 1982. La phytoécologie du Sahel et du terrain d'étude. In: F.W.T. Penning de Vries and A.M. Djitèye, Eds. La productivité des pâturages sahéliens. Une étude des sols, des végétations et de l'exploitation de cette ressource naturelle. Agric. Res. Rep. (Versl. landbouwkd. Onderz.) 918, Pudoc, Wageningen. pp. 71–83.

Cook, M.G. and L.T. Evans, 1983a. Nutrient responses of seedlings of wild and cultivated *Oryza* species. Field Crops Res. 6, 205–218.

Cook, M.G. and L.T. Evans, 1983b. Some physiological aspects of the domestication and improvement of rice (*Oryza* spp.). Field Crops Res. 6, 219–238.

Coop, I.E., 1966a. The response of ewes to flushing. W. Rev. Anim. Prod. 4, 69–75.

Coop, I.E., 1966b. Effect of flushing on reproductive performance of ewes. J. agric. Sci. (Camb.) 67, 305–323.

Coop, I.E., Ed., 1982. Sheep and goat production. Elsevier Sci. Publ. Co., Amsterdam, Neth. 492 pp.

Corbett, J.L., 1968. Variation in the yield and composition of milk of grazing Merino ewes. Aust. J. agric. Res. 19, 283–294.

Crawley, M.I., 1983. Herbivory. Blackwell, Oxford. 437 pp.

Dalling, M.J., G. Boland and J.H. Wilson, 1976. Relation between acid proteinase activity and redistribution of nitrogen during grain development in wheat. Aust. J. Plant Physiol. 3, 721–730.

Dann, P.R., 1968. Effect of clipping on yield of wheat. Aust. J. Exp. Agric. Animal Husb. 8, 731.

Degen, A.A., R.W. Benjamin and E. Eyal, 1987. A note on increasing lamb production of fat-tailed Awassi and German Mutton Merino sheep grazing in a semi-arid area. Anim. Prod. 44, 169–172.

Dobben, W.H. van, 1962. Influence of temperature and light conditions on dry-matter distribution, development rate and yield in arable crops. Neth. J. agric. Sci. 10, 377-389.

Dolman, A.J., 1987. Summer and winter rainfall interception in an oak forest: predictions with an analytical and a numerical simulation model. J. Hydrol. 90, 1–9.

Donald, C.M., 1965. The progress of Australian agriculture and the role of pastures in environmental change. Aust. J. agric. Sci. 27, 187–198.

Driel, G.J. van, C. van Ravenzwaaij, J. Spronk and F.R. Veeneklaas (1982). Objectives and potentials of the Dutch economy in the Eighties. In: M. Despontin, P. Nijkamp and J. Spronk Eds. Economic planning with conflicting goals. Springer Verlag. pp. 55–72.

Driel, G.J. van, C. van Ravenzwaaij. J. Spronk and F.R. Veeneklaas (1983). Grenzen en mogelijkheden van het economisch stelsel in Nederland (Limits and possibilities of the economic system in the Netherlands) WRR Voorstudies en Achtergronden Y 40, Staatsuitgeverij, 's Gravenhage.

Eijk, C.J. van, F.R. Veeneklaas and C.T. de Wit, 1986. Mogelijkheden voor volledige werkgelegenheid. Een verkenning van beleidsruimte. Econ. Stat. Ber. 71, 132-138. (English translation: 'A modelling approach to survey margins for policy' is available).

Epstein, H., 1970. Fettschwanzschäfe und Fettsteinschäfe. Die neue Brehambücherei. A Ziemsen Verlag, Wittenberg, Lutherstadt. 167 pp.

Epstein, H., 1971. The origin of domestic animals of Africa. Vol. II. African Publ. Corp., New York. 719 pp.

Epstein, H., 1985. The Awassi sheep with special reference to the improved dairy type. FAO, Rome. 282 pp.

Evenari, M., L. Shanan and N.H. Tadmor, 1971. The Negev- The challenge of a desert. Harvard Univ. Press, Cambridge, Mass. 345 pp.

Everett, G.C., 1966. Maternal food consumption and foetal growth in Merino sheep. Proc. Aust. Soc. Anim. Prod. 6, 91–101.

Eyal, E., R.W. Benjamin and N.H. Tadmor, 1975. Sheep production on seeded legumes, planted shrubs and dryland grain in a semi-arid region of Israel. J. Range Manage. 28, 100–107.

Eyal, E., H. Goot, W.C. Foote and D.H. Matthews, 1984. The promotion of prolific strains of sheep by nutritional and managerial means. Internal Rep., ARO-Volcani Center, Bet Dagan, Israel. 222 pp.

Farquhar, G.D., R. Wetselaar and B. Weir, 1983. Gaseous nitrogen losses from plants. In: J.R. Freney and J.R. Simpson, Eds. Gaseous loss of nitrogen from plant-soil systems. Martinus Nijhoff/Dr. W. Junk Publ., The Hague/ Boston/Lancaster. pp. 159–180.

Feigenbaum, S., N.G. Seligman and R.W. Benjamin, 1984. Fate of nitrogen-15 applied to spring wheat grown for three consecutive years in a semi-arid region. J. Soil Sci. Soc. Amer. 48, 838–843.

Fillery, I.R.P., 1983. Biological denitrification. In: J.R. Freney and J.R. Simpson, Eds. Gaseous loss of nitrogen from plant-soil systems. Martinus Nijhoff/Dr. W. Junk Publ., The Hague/ Boston/Lancaster. pp. 33–64.

Finci, M., 1957. The improvement of the Awassi breed of sheep in Israel. Weizmann Science Press of Israel, Rehovot. 106 pp.

Folman, J., E. Eyal and R. Volcani, 1966. Milk yields and weight gains of lambs in a mutton flock. J. Agric. Sci. (Camb.) 67, 369–370.

Folman, J., A. Lawi and E. Eyal, 1976. The effect of feeding an all-concentrate diet on the growth rate and feed efficiency of male lambs. Hassadeh 57, 123–126. (Hebrew.)

Ford, M.A. and G.N. Thorne, 1975. Effects of variations in temperature and light intensity at different times on the growth and yield of spring wheat. Ann. appl. Biol. 80, 283–289.

Freney, J.R., J.R. Simpson and O.T. Denmead, 1983. Volatilization of ammonia. In: J.R. Freney and J.R. Simpson, Eds. Gaseous loss of nitrogen from plant-soil systems. Martinus Nijhoff/Dr. W. Junk Publ., The Hague/ Boston/Lancaster. pp. 1–32.

Friedrich, J.W. and R.C. Huffaker, 1980. Photosynthesis, leaf resistances and ribulose-1,5-biphosphate carboxylase degradation in senescent barley leaves. Plant Physiol. 65, 1103–1107.

Gale, J. and R.M. Hagan, 1966. Plant antitranspirants. Ann. Rev. Plant Physiol. 17, 269–282.

Gallagher, J.N., 1979. Field studies of cereal leaf growth. I. Initiation and expansion in relation to temperature and ontogeny. J. exp. Bot. 30, 625–636.

Goot, H., 1986. Development of Assaf, a synthetic breed of dairy sheep in Israel. Eur. Assoc. Anim. Prod. 37th Annu. Meeting, Budapest, Hungary. 29 pp.

Goot, H., Y. Folman, R.W. Benjamin and D. Drori, 1976. Finn-Merino and Finn-Awassi crosses in the semi-arid zone of Israel. Eur. Assoc. Anim. Prod. 27th Annu. Meeting. Zürich. pp 1–7.

Goudriaan, J., 1977. Crop micrometeorology: a simulation study. Simulation Monographs, Pudoc, Wageningen. 249 pp.

Goudriaan, J., 1986. A simple and fast numerical method for the computation of daily totals of crop photosynthesis. Agric. and Forest Meteorol. 38, 249–254.

Goudriaan, J. and H. van Keulen, 1979. The direct and indirect effects of nitrogen shortage on photosynthesis and transpiration in maize and sunflower. Neth. J. agric. Sci. 27, 227–234.

Goudriaan, J. and H.H. van Laar, 1978. Calculation of daily totals of the gross CO_2 assimilation of leaf canopies. Neth. J. agric. Sci. 26, 373–382.

Graham, N. McC, and T.W. Searle, 1975. Studies of weaner sheep during and after weight stasis. I. Energy and nutrition utilization. Aust. J. agric. Res. 26, 343–353.

Greenwood, D.J., 1966. Nitrogen stress in wheat- its measurement and relation to leaf nitrogen. Plant Soil 24, 278–288.

Greenwood, D.J. and Z.V. Titmanis, 1966. The effect of age on nitrogen stress and its relation to leaf nitrogen and leaf elongation in a grass. Plant Soil 24, 379–389.

Gutman, M., 1979. Primary production of transitional mediterranean steppe. Proc. 1st. Int. Rangel. Congress, Denver, USA. pp. 225–228.

Hadjipieris, G., J.G.W. Jones, R.H. Whimble and W. Holmes, 1966. Studies on feed intake and feed utilization by sheep. II. The utilization of feed by ewes. J. agric. Sci. (Camb.) 66, 341–349.

Hadley, G., 1982. Linear programming. 3rd ed., Addison-Wesley Publishing Co. 520 pp.

Hanks, R.J., 1974. Model for predicting plant yield as influenced by water use. Agron. J. 66, 660–665.

Hanson, A.D. and W.D. Hitz, 1983. Whole-plant response to water deficits: Water deficits and the nitrogen economy. In: H.M. Taylor, W.R. Jordan and T.R. Sinclair, Eds. Limitations of efficient water use in crop production. ASA Monograph, ASA, Madison, Wisc. pp. 331-343.

Harpaz, Y., 1975. Simulation of the nitrogen balance in semi-arid regions. Ph.D. Thesis, Hebrew University, Jerusalem. 134 pp.

Hartog, J.A., G.J. van Driel and C. van Ravenzwaaij, 1979. Limits to the welfare state. Martinus Nijhof, Boston, The Hague, London. 180 pp.

Hillel, D., 1977. Computer simulation of soil-water dynamics. A compendium of recent work. Ottawa, IRDC. 214 pp.

Hochman, E., R.E. Just and D. Zilberman, 1985. The dynamics of agricultural development in sparsely populated areas: The case of the Arava. In: Y. Gradus (ed.). Desert development, D. Reidel Publ. Co. p. 256-270.

Hodge, R.W., 1966. The relative pasture intake of grazing lambs at two levels of milk intake. Aust. J. Exp. Agric. Anim. Husb. 6, 314-316.

Hodgson, J., 1979. Nomenclature and definitions in grazing studies. Grass. For. Sci. 34, 11-18.

Hoogmoed, W.B. and L. Stroosnijder, 1984. Crust formation on sandy soils in the Sahel. I. Rainfall and infiltration. Soil and Till. Res. 4, 5-23.

Ishihara, K., H. Ebara, T. Hirawasa and T. Ogura, 1978. The relationship between environmental factors and behaviour of stomata in the rice plants. VII. The relation between nitrogen content in leaf blades and stomatal aperture. Japan J. Crop Sci. 47, 664-673.

Jeffries, B.C., 1961. Body condition and its use in management. Tasmanian J. Agric. 32, 19-21.

Johnson, I.R. and A.J. Parsons, 1985. A theoretical analysis of grass growth under grazing. J. Theor. Biol. 112, 345-367.

Jones, M.B., 1963. Effect of sulfur applied and date of harvest on yield, sulfate sulfur concentration, and total sulfur uptake of five annual grassland species. Agron. J. 55, 251-254.

Jones, M.B., 1964. Effect of applied sulfur on yield and sulfur uptake of various California dryland pasture species. Agron. J. 56, 235-237.

Jones, R.J. and R.L. Sandland, 1974. The relation between animal gain and stocking rate. J. Agric. Sci. (Camb.) 83, 335-342.

Kanemasu, E.T., 1983. Yield and water-use relationships: Some problems of relating grain yield to transpiration. In: H.M. Taylor, W.R. Jordan and T.R. Sinclair, Eds. Limitations of efficient water use in crop production. ASA Monograph, ASA, Madison, Wisc. pp. 331-343.

Keulen, H. van, 1975. Simulation of water use and herbage growth in arid regions. Simulation Monographs, Pudoc, Wageningen. 176 pp.

Keulen, H. van, 1976. Evaluation of models. In: G.W. Arnold and C.T. de Wit, Eds. Critical evaluation of systems analysis in ecosystems research and management. Simulation Monographs, Pudoc, Wageningen. pp. 22-29.

Keulen, H. van, 1977. Nitrogen requirements of rice with special reference to Java. Contr. Central Res. Inst. Agric., Bogor no. 30. 67 pp.

Keulen, H. van and H.D.J. van Heemst, 1979. Crop response to the supply of macronutrients. Agric. Res. Rep. (Versl. landbouwkd. Onderz.) 916, Pudoc, Wageningen. 76 pp.

Keulen, H. van and N.G. Seligman, 1987. Simulation of water use, nitrogen nutrition and growth of a spring wheat crop. Simulation Monographs, Pudoc, Wageningen. 310 pp.

Keulen, H. van, N.G. Seligman and J. Goudriaan, 1975. Availability of anions in the growth medium to roots of an actively growing plant. Neth. J. agric. Sci. 23, 131-138.

Keulen, H. van, N.G. Seligman and R.W. Benjamin, 1981. Simulation of water use and herbage growth in arid regions. A re-evaluation and further development of the model 'Arid Crop'. Agric. Syst. 6, 159-193.

Keulen, H. van, J. Goudriaan and N.G. Seligman, 1989. Modelling the effects of nitrogen on canopy development and growth. In: G. Russell, B. Marshall and P.G. Jarvis, Eds. Plant canopies, their growth, form and fuction. SEB Seminar Series, Cambridge Univ. Press. pp. 83-104.

Kiesselbach, T.A., 1916. Transpiration as a factor in crop production. Univ. of Nebraska, Agr. Exp. Sta. Bull. 6

Kramer, Th., 1979. Yield-protein relationships in cereal varieties. In: J.H.J. Spiertz and Th. Kramer, Eds. Crop physiology and cereal breeding. Pudoc, Wageningen, The Netherlands. pp. 161–165.

Krol, O., 1978. Digestibility and methods of calculating feeds for ruminants. Bull. 202/102, Dep. Extension Services, Ministry Agric. Israel. 12 pp. (Hebrew.)

Ladd, J.N., 1981. The use of N-15 following organic matter turnover, with specific reference to rotation systems. Plant Soil 59, 410–421.

Lapins, P. and E.R. Watson, 1970. Loss of nitrogen from maturing plants. Aust. J. exp. Agric. Anim. Husb. 10, 599–603.

Large, R.V. and P.D. Penning, 1967. The artificial rearing of lambs on cold reconstituted whole milk and on milk substitute. J. agric. Sci. (Camb.) 69, 405–409.

Le Houérou, H.N. and C.H. Hoste, 1977. Rangeland production and annual rainfall relations in the Mediterranean Basin and in the African Sahelo-Sudanian zone. J. Range Manage. 30, 181–189.

Le Houérou, H.N., R.L. Bingham and W. Skerbek, 1988. Relationship between the variability of primary production and the variability of annual precipitation in world arid lands. J. arid Env. 15, 1–18.

Lof, H., 1976. Water use efficiency and competition between arid zone annuals, especially the grasses Phalaris minor and Hordeum murinum. Agric. Res. Rep. (Versl. landbouwkd. Onderz.) 853, Pudoc, Wageningen. 109 pp.

Lomas, J., 1972. Forecasting wheat yields from rainfall data in Iran. WMO bulletin, Jan. 1972.

Lomas, J. and Y. Shashoua, 1973. The effect of rainfall on wheat yields in an arid region. In: R.O. Slatyer, Ed. Plant response to climatic factors, Proc. of the Uppsala Symp., UNESCO, Paris. pp. 531–537.

Lomas, J. and Y. Shashoua, 1974. The dependence of wheat yields and grain weight in a semi-arid region on rainfall and the number of hot dry days. Isr. J. agric. Res. 23, 113–121.

Loria, M. and I. Noy-Meir, 1980. Dynamics of some annual populations in a desert loess plain. Israel J. Bot. 28, 211–225.

Mason, I.L., 1951. A world dictionary of breeds, types and varieties of livestock. Tech. Comm. 8, Commonw. Agric. Bur., Farnham Royal, UK. 268 pp.

Mason, I.L., 1967. The sheep breeds of the Mediterranean. Commonw. Agric. Bur., Farnham Royal, UK. 215 pp.

McNaughton, S.J., 1979. Grazing as an optimization process: Grass-ungulate relationships in the Serengeti. Arch. Nat. 113, 691–703.

McNeal, F.H., M.A. Berg and C.A. Watson, 1966. Nitrogen and dry matter in five spring wheat varieties at successive stages of development. Agron. J. 58, 605–608.

Monteith, J.L., 1965. Light and crop productioon. Field Crop Abstr. 4, 213–219.

Moore, R.M., Ed., 1970. Australian grasslands. Austral. Nat. Univ. Press, Canberra. 455 pp.

Morag, M., A.A. Degen and E. Popliker, 1973. The reproductive performance of German Mutton Merino ewes in a hot arid climate. Z. Tierz. Züchtungsbiol. 89, 340–345.

Morin, J. and Y. Benyamini, 1977. Rainfall infiltration in bare soils. Water Res. Res. 13, 813–817.

Morley, F.H.W. and C.R.W. Spedding, 1968. Agricultural systems and grazing experiments. Herb. Abstr. 38, 279–287.

Mott, G.O., 1960. Grazing pressure and the measurement of pasture production. Proc. 8th Inter. Grassl. Congr. pp. 606–611.

Nair, T.V.R., H.L. Grover and Y.P. Abrol, 1978. Nitrogen metabolism of the upper three leaf blades of wheat at different soil nitrogen levels. II. Protease activity and mobilization of reduced nitrogen to the developing grains. Physiol. Plant. 42, 293–300.

Neales, T.F. and L.D. Incoll, 1968. The control of leaf photosynthesis rate by the level of assimilate concentration in the leaf. A review of the hypothesis. Bot. Rev. 34, 107–125.

Noy-Meir, I., 1975a. Stability of grazing systems: An application of predator-prey graphs. J. Ecol. 63, 459-481.

Noy-Meir, I., 1975b. Primary and secondary production in sedentary and nomadic grazing systems in the semi-arid region: Analysis and modelling. Res. Rep. Ford Foundation. 367 pp.

Noy-Meir, I., 1976. Rotational grazing in a continuously growing pasture: A simple model. Agric. Systems 1, 87-112.

Noy-Meir, I., 1978. Grazing and production in seasonal pastures: Analysis of a simple model. J. Appl. Ecol. 15, 809-835.

Noy-Meir, I. and Y. Harpaz, 1977. Agro-ecosystems in Israel. Agro-ecosystems 4, 143-167.

Noy-Meir, I. and N.G. Seligman, 1979. Management of semi-arid ecosystems in Israel. In: B. Walker, Ed. Management of semi-arid ecosystems, Elsevier Sci. Publ. Co., Amsterdam. pp. 113-160.

NRC, 1975. Nutrient requirements of domestic animals. 5. Nutrient requirements of sheep. Nat. Acad. Sci. Washington. 72 pp.

NRC, 1985. Nutrient requirements of sheep. 6. Nat. Acad. Sci. Washington. 97 pp.

Ofer, J., 1980. The ecology of ant populations of the genus Messor and their influence on the soil and flora in pasture. Ph.D. Thesis, Hebrew Univ. of Jerusalem, Jerusalem. 204 pp.

Ofer, Y., R.W. Benjamin and N.H. Tadmor, 1967. Fertilizer application to arid pasture in a dry year (1965/66). Div. of Scientific Publ., Volcani Institute, Bet Dagan, Israel, Prel. Rep. No. 565, Proj. No. 17/01003. 10 pp. (Hebrew, with English summary.)

Os, A.J. van, 1967. The influence of nitrogen supply on the distribution of dry matter in spring rye. Jaarb. Inst. Biol. Scheik. Onderz. Landbouwgewassen 1966, 51-65.

Parnas, H., 1975. Model for decomposition of organic material by micro-organisms. Soil Biol. Biochem. 7, 161-169.

Pearman, I., S.M. Thomas and G.N. Thorne, 1981. Dark respiration of several varieties of winter wheat given different amounts of nitrogen fertilizer. Ann. Bot. 47, 535-546.

Peart, J.N., 1968. Some effects of liveweight and body condition on the milk production of Blackface ewes. J. agric. Sci. (Camb.) 70, 331-338.

Penman, H.L., 1948. Natural evaporation from open water, bare soil and grass. Proc. R. Soc. A193, 120-145.

Penman, H.L., 1963. Vegetation and Hydrology. Techn. Comm. 53, Commonwealth Bureau of Soils, Harpenden. 124 pp.

Penning de Vries, F.W.T., 1974. Substrate utilization and respiration in relation to growth and maintenance in higher plants. Neth. J. agric. Sci. 22, 40-44.

Penning de Vries, F.W.T., 1975. The cost of maintenance processes in plant cells. Ann. Bot. 39, 77-92.

Penning de Vries, F.W.T., 1982. Le potentiel physiologique des pâturages et des cultures agricoles. In: F.W.T. Penning de Vries and A.M. Djitèye, Eds. La productivité des pâturages sahéliens. Une étude des sols, des végétations et de l'exploitation de cette ressource naturelle. Agric. Res. Rep. (Versl. landbouwkd. Onderz.) 918, Pudoc, Wageningen. pp. 87-98.

Penning de Vries, F.W.T. and A. M. Djitèye (Eds.), 1982. La productivité des pâturages sahéliens. Une étude des sols, des végétations et de l'exploitation de cette ressource naturelle. Agric. Res. Rep. (Versl. landbouwkd. Onderz.) 918, Pudoc, Wageningen. 525 pp.

Penning de Vries, F.W.T. and H. van Keulen, 1982. La production actuelle et l'action de l'azote et du phosphore. In: F.W.T. Penning de Vries and A.M. Djitèye, Eds. La productivité des pâturages sahéliens. Une étude des sols, des végétations et de l'exploitation de cette ressource naturelle. Agric. Res. Rep. (Versl. landbouwkd. Onderz.) 918, Pudoc, Wageningen. pp. 196-226.

Penning de Vries, F.W.T., J.M. Witlage and D. Kremer, 1979. Rates of respiration and of increase in structural dry weight in young wheat, ryegrass and maize plants in relation to temperature, to water stress and to their sugar content. Ann. Bot. 44, 595-609.

Power, J.F., 1980a. Response of semi-arid grassland sites to nitrogen fertilization. I. Plant growth and water use. Soil Sci. Soc. Am. Proc. 44, 545-550.

Power, J.F., 1980b. Response of semi-arid grassland sites to nitrogen fertilization. II. Fertilizer recovery. Soil Sci. Soc. Am. Proc. 44, 550-555.

Radin, J.W., 1983. Control of plant growth by nitrogen: difference between cereals and broadleaf species. Plant. Cell. Env. 6, 65–68.

Radin, J.W. and R.C. Ackerson, 1981. Water relations of cotton plants under nitrogen deficiency. III. Stomatal conductance, photosynthesis, and abscissic acid accumulation during drought. Plant Physiol. 67, 115–119.

Radin, J.W. and Boyer, J.S., 1982. Control of leaf expansion by nitrogen nutrition in sunflower plants: role of hydraulic conductivity and turgor. Plant Physiol. 69, 771–775.

Radin, J.W. and L.L. Parker, 1979a. Water relations of cotton plants under nitrogen deficiency. I. Dependence upon leaf structure. Plant Physiol. 64, 495–498.

Radin, J.W. and L.L. Parker, 1979b. Water relations of cotton plants under nitrogen deficiency. II. Environmental interactions on stomata. Plant Physiol. 64, 499–501.

Reij, C., P. Mulder and L. Begeman, 1988. Water harvesting for plant production. World Bank Techn. Paper no. 91. 123 pp.

Ridder, N. de, N.G. Seligman and H. van Keulen, 1981. Analysis of environmental and species effects on the magnitude of biomass investment in the reproductive effort of annual pasture plants. Oecologia (Berlin) 49, 263–271.

Ridder, N. de, R.W. Benjamin and H. van Keulen, 1986. Forage selection and performance of sheep grazing dry annual range. J. Arid Environ. 10, 39–51.

Rietveld, J.J., 1978. Soil non-wettability and its relevance as a contributing factor to surface runoff on sandy dune soils in Mali. Intern. Rep. Dept. Theoretical Prod. Ecol., Agric. Univ., Wageningen. 179 pp.

Rijtema, P.E., 1965. An analysis of actual evapotranspiration. Agric. Res. Rep. (Versl. landbouwkd. Onderz.) 659, Pudoc, Wageningen. 107 pp.

Ritchie, J.R., 1972. Model for predicting evaporation from a row crop with incomplete soil cover. Water Res. Res. 8, 1204–1213.

Russell, E.J., 1950. Soil conditions and plant growth. Longmans, Green and Co., London. 635 pp.

Seginer, I., and Y. Morin, 1970. A model of surface crusting and infiltration of bare soils. Water Res. Res. 6, 629-633.

Seligman, N.G. and H. van Keulen, 1981. PAPRAN: A simulation model of annual pasture production limited by rainfall and nitrogen. In: M.J. Frissel and J.A.van Veen, Eds. Simulation of nitrogen behaviour of soil-plant systems. Pudoc, Wageningen. pp. 192–220.

Seligman, N.G. and H. van Keulen, 1989. Herbage production of a mediterranean grassland in relation to soil depth, rainfall and nitrogen nutrition: A simulation study. Ecol. Modelling 47, 303–311.

Seligman, N.G., Z. Rosenzaft, N.H. Tadmor, J. Katznelson and Z. Naveh, 1960. Natural pastures of Israel. Sifriat Hapoalim, Merhavia, Israel. 378 pp. (Hebrew with English summary.)

Seligman, N.G., H.van Keulen, A. Yulzari, R.Yonathan and R.W. Benjamin, 1976. The effect of abundant nitrogen fertilizer application on the seasonal change in mineral concentration in annual mediterranean pasture species. Prel. Report no. 754, Div. Sci. Publ., Volcani Center, P.O.Box 6, Bet Dagan, Israel.

Seligman, N.G., R.W. Benjamin and E. Eyal, 1981. Migda System 1 (MIGS1): A model for studying management systems of an integrated sheep-wheat farm in the semi-arid zome of Israel. Spec. Publ. 207, Division Sci. Publ. Volcani Center, Bet Dagan.

Seligman, N.G., R.W. Benjamin and M. Yanuka, 1981. Early season defoliation, seedling density and production of some improved barley and spring wheat cultivars growing in a semi-arid temperate winder-rainfall region. Internal Rep. 105-E, Dep. Range Sci., Agric. Res. Organization, Bet Dagan.

Seligman, N.G., R.S. Loomis, R.S. Burke and A. Abshahi, 1983. Nitrogen nutrition and phenological development in field-grown wheat. J. agric. Sci. (Camb.) 101, 691–697.

Seligman, N.G., S. Feigenbaum, R.W. Benjamin and D. Feinerman, 1985. Efficiency of a fallow as a store for fertilizer nitrogen in a semi-arid region. J. agric. Sci. (Camb.) 105, 245–249.

Seligman, N.G., S. Feigenbaum, D. Feinerman and R.W. Benjamin, 1986. Uptake of nitrogen from high C-to-N ratio, ^{15}N-labelled organic residues by spring wheat grown under semi-arid conditions. Soil Biol. Biochem. 18, 303–307.

Shanan, L., M. Evenari and N.H. Tadmor, 1967. Rainfall patterns in the central Negev desert. Isr. Expl. J. 17, 163–184.

Shimshi, D. and U. Kafkafi, 1978. The effect of supplementary irrigation and nitrogen fertilisation on wheat (*Triticum aestivum* L.). Irr. Sci. 1, 27–38.

Simon, H.A., (1955). A behavioural model of rational choice. Quarterly J. Econ. 69, 99–118.

Sinclair, T.R. and T. Horie, 1989. Leaf nitrogen, photosynthesis, and crop radiation use efficiency: A review. Crop Sci. 29, 90–98.

Sinclair, T.R. and C.T. de Wit, 1976. Analysis of carbon and nitrogen limitation to soybean yield. Agron. J. 68, 319-324.

Slatyer, R.O., 1967. Plant-water relationships. Acad. Press, London.

Sofield, I., I.F. Wardlaw, L.T. Evans and S.Y. Zee, 1977. Nitrogen, phosphorus and water contents during grain development and maturation in wheat. Aust. J. Plant Physiol. 4, 799–810.

Spedding, C.R.W., 1970. Sheep production and grazing management. Bailliere. Tindall and Cassel, London. 380 pp.

Spedding, C.R.W., 1975. The biology of agricultural systems. Acad. Press, London. 261 pp.

Spronk, J. and F.R. Veeneklaas (1983). A feasability study of economic and environmental scenarios by means of Interactive Multiple-Goal Programming. Regional Sci. and Urban Econ. 13, 141–160.

Stapper, M., 1984. SIMTAG, a simulation model of wheat genotypes. Univ. of New England, ICARDA. 108 pp.

Stroosnijder, L., 1982. Simulation of the soil water balance. In: F.W.T. Penning de Vries and A.M. Djitèye, Eds. La productivité des pâturages sahéliens. Une étude des sols, des végétations et de l'exploitation de cette ressource naturelle. Agric. Res. Rep. (Versl. landbouwkd. Onderz.) 918, Pudoc, Wageningen. pp. 175–193.

Stroosnijder, L. and D. Koné, 1982. Le bilan d'eau du sol. In: F.W.T. Penning de Vries and A.M. Djitèye, Eds. La productivité des pâturages sahéliens. Une étude des sols, des végétations et de l'exploitation de cette ressource naturelle. Agric. Res. Rep. (Versl. landbouwkd. Onderz.) 918, Pudoc, Wageningen. pp. 133–165.

Sunderland, N., 1960. Cell division and expansion in the growth of the leaf. J. exp. Bot. 11, 68–80.

Tadmor, N.H., B. Yogev, E. Eyal and L. Shanan, 1966. Seeded dryland range in the northern Negev of Israel (Third progress report, 1964/65). Div. of Scientific Publ., Volcani Institute, Bet Dagan, Israel, Prelim. Rep. No. 545, Proj. no. 17/01003. 84 pp. (Hebrew, with English summary.)

Tadmor, N.H., E. Eyal, R.W. Benjamin and L. Shanan, 1967. Seeded dryland range in the Northern Negev of Israel. (Fourth Progress Report, 1965/66). Div. of Scientific Publ., Volcani Institute, Bet Dagan, Israel, Prelim. Rep. no. 581, Proj. no. 291/01003. 95 pp. (Hebrew with English summary.)

Tadmor, N.H., E. Eyal and R.W. Benjamin, 1970. Seeded dryland range in the Northern Negev of Israel. (Fifth Progress Report, 1966/67). Div. of Scientific Publ., Volcani Institute, Bet Dagan, Israel, Prelim. Rep. no. 678, Proj. no. 291/01003. 105 pp. (Hebrew with English summary.)

Tadmor, N.H., E. Eyal and R.W. Benjamin, 1974. Plant and sheep production on semi-arid annual grassland in Israel. J. Range Manage. 27, 427–432.

Tanner, C.B. and T.R. Sinclair, 1983. Efficient water use in crop production: Research or research. In: H.M. Taylor, W.R. Jordan and T.R. Sinclair, Eds. Limitations of efficient water use in crop production. ASA Monograph, ASA, Madison, Wisc. pp. 1–27.

Theron, J.J., 1951. The influence of plants on the mineralization of nitrogen and the maintenance of organic matter in the soil. J. agric. Sci. (Camb.) 41, 289–296.

Theron, J.J. and D.G. Haylett, 1953. The regeneration of soil humus under a grass ley. Emp. J. exptl. Agric. 2, 80–98.

Treacher, T.T., 1970. Effect of nutrition in late pregnancy on subsequent milk production in ewes. Anim. Prod. 12, 23–36.

Ungar, E.D., 1984. Management of agro-pastoral systems in a semi-arid region. Ph. D. Thesis, Hebrew Univ. Jerusalem. 168 pp.

210

Ungar, E.D., 1990. Management of agropastoral systems in a semi-arid region. Simulation Monographs, Pudoc, Wageningen. 221 pp.

Ungar, E.D. and I. Noy-Meir, 1988. Herbage intake in relation to availability and sward structure: Grazing processes and optimal foraging. J. Appl. Ecol. 25, 1045-1062.

Veeneklaas, F.R., 1990. Dovetailing technical and economical analysis. Ph.D. thesis Erasmus University, Rotterdam. 159 pp.

Versteeg, M.N., 1985. Factors influencing the productivity of irrigated crops in Southern Peru, in relation to prediction by simulation models. Pudoc, Wageningen. 182 pp.

Viets, F.G. Jr., 1962. Fertilizers and the efficient use of water. Adv. Agron. 14, 223–264.

Vos, J., 1981. Effects of temperature and nitrogen supply on post-floral growth of wheat; measurements and simulations. Agric. Res. Rep. (Versl. landbouwkd. Onderz.) 911. Pudoc, Wageningen. 164 pp.

Walter, H. and O.H. Volk, 1954. Grundlagen der Weiderwirtschaft in Süd West Afrika. Stuttgart, Ulmer Verl. 218 pp.

Westoby, M., 1980. Elements of a theory of vegetation dynamics in arid rangelands. Israel. J. Bot. 28, 169–194.

Williams, W.A., R.M. Love and J.P. Conrad, 1956. Range improvement in California by seeding annual clovers, fertilization and grazing management. J. Range Manage. 9, 28–33.

Willoughby, W.M., 1959. Limitations to animal production imposed by seasonal fluctuations in pasture and management procedures. Aust. J. agric. Res. 10, 248–268.

Wilson, J.R., 1975a. Influence of temperature and nitrogen on growth, photosynthesis and accumulation of non-structural carbohydrate in a tropical grass, *Panicum maximum* var. *trichoglume*. Neth. J. agric. Sci. 23, 48–61.

Wilson, J.R., 1975b. Comparative response to nitrogen deficiency of a tropical and a temperate grass in the interrelation between photosynthesis, growth and the accumulation of non-structural carbohydrate. Neth. J. agric. Sci. 23, 104–112.

Wit, C.T. de, 1953. A physical theory on placement of fertilizers. Agric. Res. Rep. (Versl. landbouwkd. Onderz.) 59.4, Staatsdrukkerij, 's-Gravenhage. 71 pp.

Wit, C.T. de, 1958. Transpiration and crop yields. Agric. Res. Rep. (Versl. landbouwkd. Onderz.) 64.4, Staatsdrukkerij, Den Haag. 88 pp.

Wit, C.T. de, 1965. Photosynthesis of leaf canopies. Agric. Res. Rep. (Versl. landbouwkd. Onderz.) 663, Pudoc, Wageningen. 57 pp.

Wit, C.T. de and H. van Keulen, 1972. Simulation of transport processes in soils. Simulation Monographs, Pudoc, Wageningen. 100 pp.

Wit., C.T. de and J.M. Krul, 1982. La production actuelle dans une situation d'équilibre. In: F.W.T. Penning de Vries and A.M. Djit]eye, Eds. La productivité des pâturages sahéliens. Une étude des sols, des végétations et de l'exploitation de cette ressource naturelle. Agric. Res. Rep. (Versl. landbouwkd. Onderz.) 918, Pudoc, Wageningen. pp. 275–283.

Wit, C.T. de *et al.*, 1978. Simulation of assimilation, respiration and transpiration of crops. Simulation Monographs, Pudoc, Wageningen. 141 pp.

Wit, C.T. de, H. Huisman and R. Rabbinge, 1987. Agriculture and the environment: Are there other ways? Agric. Systems 23, 211–236.

Wit, C.T. de, H. van Keulen, N.G. Seligman and I. Spharim, 1988. Application of interactive multiple-goal programming techniques for analysis and planning of regional agricultural development. Agric. Syst. 26, 211-230.

Wong, S.C., I.R. Cowan and G.D. Farquhar, 1979. Stomatal conductance correlates with photosynthetic capacity. Nature 282, 424–426.

Yanuka, M., N.G. Seligman and R.W. Benjamin, 1981. The effect of defoliation early in the growing season on the production of some improved barley and wheat cultivars in a semi-arid region of Israel. Internal Report, Dept. of Range Sci. Agric. Res. Organization, Bet Dagan, Israel (mimeographed).

Yoshida, S. and V. Coronel, 1976. Nitrogen nutrition, leaf resistance and leaf photosynthetic rate of the rice plant. Soil Sci. Plant Nutr. 22, 207–211.

Yoshida, S. and Y. Hayakawa, 1970. Effects of mineral nutrition on tillering of rice. Soil Sci. Plant Nutr. 16, 186–191.

Zaban, H., 1981. A study to determine the optimal rainfed land-use systems in a semi-arid region of Israel. Ph. D. Thesis, Univ. of Reading. 197 pp.

Systems Approaches for Sustainable Agricultural Development

1. Th. Alberda, H. van Keulen, N.G. Seligman and C.T. de Wit (eds.): *Food from Dry Lands*. An Integrated Approach to Planning of Agricultural Development. 1992 ISBN 0-7923-1877-3

2. F.W.T. Penning de Vries, P.S. Teng and K. Metselaar (eds.): *Systems Approaches for Agricultural Development*. Proceedings of the International Symposium on Systems Approaches for Agricultural Development, Bangkok, Thailand, 2–6 December (1991). 1992
ISBN Hb 0-7923-1880-3; Pb 0-7923-1881-1

KLUWER ACADEMIC PUBLISHERS – DORDRECHT / BOSTON / LONDON

The manufacturer's authorised representative in the EU is Springer
Nature Customer Service Centre GmbH, Europaplatz 3, 69115 Heidelberg,
Germany. If you have any concerns regarding our products, please
contact ProductSafety@springernature.com

Printed and bound by CPI Group (UK) Ltd, Croydon, CR0 4YY

23/04/2026
02095629-0010